ECONOMIC COMMISSION FOR EUROPE

GENEVA

SUSTAINABLE ENERGY DEVELOPMENTS IN EUROPE AND NORTH AMERICA

ECE ENERGY SERIES No. 6

UNITED NATIONS

NEW YORK, 1991

NOTE

The designations employed and the presentation of the material in this publication do not imply the expression of any opinion whatsoever on the part of the Secretariat of the United Nations concerning the legal status of any country, territory, city or area, or of its authorities, or concerning the delimitation of its frontiers or boundaries.

*
* *

Symbols of United Nations documents are composed of capital letters combined with figures. Mention of such a symbol indicates a reference to a United Nations document.

UNITED NATIONS PUBLICATION
Sales No. E.91.II.E.2
ISBN 92-1-116499-0

06000P

PREFACE

Europe and North America account for 70% of world energy consumption, 61% of which are fossil fuels. Energy trends and patterns in this region, if pursued, would heavily impact on region- and world-wide energy and ecosystems.

Would projected trends and supply structures be "sustainable", i.e. "meet the needs of the present without compromising the ability of future generations to meet their own needs" (World Commission on Environment and Development)? What adaptations are warranted? What role could and should be played by regional energy and environmental co-operation, including through the United Nations Economic Commission for Europe?

These are the issues dealt with by the present study mandated by the Senior Advisers to ECE Governments on Energy - a principal subsidiary body of the UN-ECE - in 1988 and reviewed in 1990. In line with a decision taken by the Senior Advisers to ECE Governments on Energy at their seventh session in October 1990, the study is released to the public domain, as customary under the responsibility of the secretariat.

GE.91-30023/2790B

TABLE OF CONTENTS

PART I

STUDY ON THE INTERRELATIONSHIPS BETWEEN ENVIRONMENTAL AND ENERGY POLICIES IN EUROPE AND NORTH AMERICA TILL 2010 AND BEYOND

PART II

RESEARCH NOTES

CONTENTS (<u>continued</u>)

PART II (<u>continued</u>)

PART III

THE ENERGY PROGRAMME OF UN-ECE

PART I

STUDY ON THE INTERRELATIONSHIPS BETWEEN ENVIRONMENTAL
AND ENERGY POLICIES IN EUROPE AND NORTH AMERICA TILL
2010 AND BEYOND

INTRODUCTION

A.   MANDATE

1.   Responding to concerns raised _inter alia_ by the report of the World
Commission on Environment and Development 1/ on the long-term sustainability
of world energy developments, the Senior Advisers to ECE Governments on Energy
included, at their sixth session, a study on "Interrelationships between
Environmental and Energy Policies" in the Programme of Work for 1988-1992
(ECE/ENERGY/13, Annex I, programme element 1.2). The purpose of this study
was "to review the interrelationships between energy and environment, with
particular reference to the long-term implications for energy supply and
security, and energy costs in the ECE region of environmental and climatic
considerations, including those raised in the report of the World Commission
on Environment and Development". The Senior Advisers also agreed that the
study could be elaborated with the help of rapporteurs nominated by interested
Governments.

2.   In keeping with this mandate, a Preparatory Meeting on the Study of the
Interrelationships between Environmental and Energy Policies was held in
Geneva on 17 and 18 May 1989 (ENERGY/AC.10/2). After a discussion of the
basic issues involved, the Meeting determined the objectives, breakdown and
work methods of the study. It requested the secretariat to prepare a draft of
the study, for consideration by an _Ad hoc_ meeting, which was held in Geneva
from 29 to 31 January 1990 (ENERGY/AC.10/4). It reviewed and approved the
main document (ENERGY/AC.10/R.2) as amended and took note of its addenda 1
to 8.

3.   The Senior Advisers to ECE Governments on Energy considered the study
at their seventh session (Geneva, 29 October to 1 November 1990). After
discussion, they agreed to release the study to the public domain, as customary
under the responsibility of the secretariat (ECE/ENERGY/15, para. 14 (a)).

B.   OBJECTIVES

4.   The present study assesses the interface between environmental and energy
policies by:

   (i)   developing and analysing projections of energy demand and supply to
         2000 and 2010, and evaluating longer-term scenarios elaborated in
         other fora;

  (ii)   analysing the feasibility and sustainability of these projections
         and scenarios from a long-term and region-wide standpoint;

 (iii)   describing policy responses as implemented or envisaged by ECE
         Governments;

  (iv)   identifying and discussing the issues relevant to the process of
         adaptation to environmental concerns; and

   (v)   defining the role of regional energy and environmental co-operation
         through ECE in furthering this process of adaptation.

C. SUMMARY

(a) Chapter I

5.   Under the impact of growing environmental tensions, the integration of energy and environmental policies has become a central requirement in all ECE countries. The aim is to subsume fragmented approaches (data collection, model building, systems analysis, policy formulation) under overriding societal goals and values so as to prevent, reduce or mitigate constraints. This integration, presently in its infancy, will not materialize automatically, but requires the overcoming of institutional compartmentalization, policy incompatibilities and analytical fragmentation. These requirements apply also to the international level including ECE.

(b) Chapter II

6.   One of the conditions for a successful integration of energy and environmental policies is clarity about (energy, environmental) trends, so as to identify constraints. This is why Chapter II describes energy demand and supply projections for the ECE region and its major groupings for the period 1985-2010 on the basis of assumptions concerning demographic and economic growth and oil price development.

7.   Chapter II concludes that despite a significant reduction of the amount of energy needed to produce a unit of gross domestic product, primary energy demand would continue to grow in the market economies and central and east European countries, particularly in the latter. Growth rates decline, though. Demand growth would be satisfied for the main by rising efficiency of energy use, followed by production growth; coal supply is projected to supply the biggest increment, followed by gas and nuclear power; ECE oil production would fall; new and renewable sources would make a growing contribution in absolute terms but remain secondary in relative terms. Total supplies (production and net trade) would continue to be dominated by fossil fuels (1985: 89%; 2010: 83-84%). Oil would lose market shares but still remain the No. 1 fuel, followed closely by coal and next gas and nuclear power. The energy import ratio of the market economies would not deviate much from its 1985 level, but would see its oil and gas contents increasing. The east European economies would substantially reduce energy exports, but remain net exporters; the size of net exports would not so much depend on resources, but on profitability, and participation of importers in the investment efforts of exporters.

(c) Chapter III

8.   Integrated energy and environmental policies pre-suppose an integrated assessment. This is why Chapter III evaluates the long-term sustainability of the above projections on the basis of eleven criteria reflecting macro-economic, energy, environmental and societal concerns. However deficient this approach may be (it considers the criteria in isolation rather than in their interaction), it suggests that:

(i) in the short- and medium-term till about 2010, the projected energy developments appear feasible, efficient, secure and viable options for Governments and businesses; their environmental impact could be controlled if the appropriate measures were taken. The projected scenarios appear responsive to growing demand and emerging macro-economic and societal trends towards service and information based economies. They appear compatible with reserves (which world-wide are vast and increasing) and with energy policies priorities. However, drawbacks exist and would gain importance: higher energy costs, reduced systems' reliance against temporary and localized stresses, higher import dependence on non-ECE oil suppliers, a contraction of East-West energy trade; deficient international, particularly East-West technology transfer; a slow penetration of renewable sources, and uncertainty about the scope of nuclear power growth. However, these drawbacks are not considered to jeopardize the genuine sustainability of the projected energy paths in the short- and medium-term till about 2010.

(ii) However, growingly and _further into the twenty-first century_, scenarios of the type presented are not considered "sustainable", both on account of the risk of climate variation resulting from the heavy fossil fuel bias of the projected energy paths, and also on account of the considerable drain on ECE and world conventional oil, uranium and gas reserves.

(iii) _Adapted policies are, thus, warranted_.

(d) _Chapter IV_

9. Chapter IV reviews _remedial measures_ individually (strengthening of market mechanisms, enhanced energy demand and energy supply management) and combined in "preferred policy mixes" focusing on greater efficiency in the utilization of energy, substitution of fossil fuels by renewables and nuclear energy, and of coal and oil by gas, and development of CO_2-benign and CO_2-control technologies. Considering global warming as the overriding long-term constraint, a methodology has been tested (Add. 2) to contribute to a global energy response strategy. While the merit of the exercise lies perhaps more in testing an analytical tool and understanding the interplay of influencing factors, its message is encouraging: an evolutionary, balanced and global energy response strategy could be designed to gain time for achieving longer term energy sustainability.

(e) _Chapter V_

10. Chapter V holds that integrated energy and environmental policies need a strong international dimension, which in turn requires the co-operation of the various international bodies concerned and the co-ordination and integration of their work programmes. ECE, whose member States accounted for 68% of world energy consumption in 1985 and may still account for as much as 60% in 2010, assumes a particularly important role in this process. Accordingly, the study identifies issues which require special attention.

CHAPTER I

THE INTERFACE BETWEEN ENVIRONMENTAL AND ENERGY POLICIES:
THE BASIC ISSUES

A.   INTENDED AND ACTUAL INTEGRATION, SO FAR

11. In the 1990s, ECE Governments foresee a greater commitment to
environmental concerns, the declared intent being the integration of
environmental and energy policies. As a result, environmental protection
moves from a peripheral to a more central role in the formulation of energy
policies. Ecological sustainability becomes a yardstick against which
suggested energy activities are and will be evaluated in the same vein as
concerns of energy security or service.

12. At the same time the concept of "environment" has broadened. It is no
longer restricted to the short-term environmental impact of a given energy
facility, but extends into the complex interface between energy demand and
supply, natural resources and society, including the international community.
Also concerns stretch beyond traditional pollutants such as dust, noise,
sulphur dioxide, nitrogen oxides, heavy metals, volatile organic compounds,
radionuclides and, whether or not a pollutant: CO_2.

13. Measured against the intentions, the integration of energy and
environmental policies is still in its infancy. Policies are often
contradictory rather than complementary; institutional structures remain
compartmentalized rather than multi-disciplinary; measures continue to be
taken on an ad hoc basis rather than systematically and are, more often than
not, curative rather than preventive.

B.   EXPECTATIONS

14. Are these difficulties temporary, likely to be overcome? Or are they
indicative of basic incompatibilities?

15. A historical observation may help to answer these questions. Concerns
about ecological disruptions gained particular political strength in the more
affluent ECE countries: the more affluent a society, the greater the pressure
on its human and natural resources, the more pressing the need to preserve the
environment. This cause-effect relationship will also apply to the future and
to other countries as long as ECE countries progress on the road to economic
welfare. The integration of energy and environmental policies appears,
therefore, as a lasting necessity of growing urgency.

16. This integration will not materialize by itself however. It requires an
adaptation of policy formulation and a concept which transcends fragmented and
contradictory policies.

C.   GOALS

17. At the national level, energy and environmental policy advisers need to
overcome compartmentalized government structures and to ensure their early and
full participation in the definition of objectives, the identification of
priorities, the consideration of alternative options and in the implementation
and monitoring of policies. There is also a need for environmental and energy

analysts to bridge differences in the data-bases, modelling approaches and systems analysis so as to establish a common statistical and analytical basis on which to build a common policy. There is lastly a need for establishing a mechanism for arbitrating conflicting views. 2/

18. A similar approach needs to be applied also at the international level. In ECE, several of the Principal Subsidiary Bodies of the Commission (Senior Advisers to ECE Governments on Energy, Senior Advisers to ECE Governments on Environmental and Water Problems, 3/ Senior Advisers to ECE Governments on Science and Technology, the Coal Committee, the Committee on Gas, the Committee on Electric Power, the Conference of European Statisticians) would need to be involved in any effort of integrating programmes on the basis of a common concept, should this prove useful.

D. CONSTRAINTS AND MEANS

19. A common concept is, indeed, paramount. The integration of fragmented policies and work methods requires a concept transcending the more narrow approaches of the various policies to be integrated. Its overriding objective would be to prevent, reduce or mitigate those constraints which ultimately limit or impede the achievement of societal goals. Constraints could be the limited absorptive capacity of an ecosystem, the vulnerability of an energy system in the case of trade restrictions, the lack of capital to finance environmental control techniques, etc.

20. Measures to cope with such constraints or "scarce goods" are numerous. They include increasing their prices, rationing their distribution, attributing them to a preferential use, enhancing the efficiency of their use, spreading their use over time (peak-shaving) or space (re-location of plants), developing substitutes, pooling resources.

21. In the international context, Governments define these overriding objectives, ultimate constraints and subsequent policy measures in accordance with their own social and economic systems. The resulting international incompatibility of national concepts emerges as a possible impediment to the international integration of energy and environmental policies. In the ECE region it is the role of ECE to build a consensus with a view to achieving common objectives.

CHAPTER II

TRENDS AND PROJECTIONS OF ENERGY DEVELOPMENTS
IN THE ECE REGION UNTIL 2010 AND BEYOND

A. NATURE OF PROJECTIONS, SCENARIOS

22. A possible basis for judgements of the long-term sustainability of energy developments in the ECE region are projections or scenarios. These are not forecasts of what is likely, or what ought to happen. More modestly, they are projections of demand/supply trends that could happen if the underlying assumptions became true. However "realistic" or "desirable" they may be, their primary purpose is the early identification of risks or opportunities associated with these energy "futures".

B.   MAIN ASSUMPTIONS

23.   While energy demand and supply depend on numerous factors, the following
factors have been considered as particularly important in the present study:
demographic and economic growth, and the development of oil prices
(see table 1).

24.   Population growth is assumed to continue, albeit at declining growth
rates. By 2010, total population in the ECE region is projected to exceed
the 1985 level by 15.4%.

25.   Two variants of economic growth have been retained, a "moderate economic
growth scenario" and a "low economic growth scenario". The first assumes a
gradual improvement of balance of payment deficits and surpluses worldwide,
including a reduction of the debt position of many developing countries;
a smooth and sustained development of trade in basic products and raw
materials; a stable expansion of global financial systems; growing world
trade; and enhanced international division of labour. The second scenario
assumes that these improvements would not or only partially occur.

26.   Both scenarios assume continued economic growth to 2000, however at
declining rates thereafter. For the market economies in the moderate growth
scenario, real GDP growth is assumed to fall from a 2.7% annual compound rate
in 1985-2000 to 2.1% in 2000-2010; in the low growth scenario, the decline is
from 2.1% to 1.4%. For the east European economies, the moderate growth
scenario assumes growth rates of Net Material Product (NMP) of 3.5% per annum
during 1985-2000 and of 2.9% during 2000-2010; in the low growth scenario, the
corresponding rates are 2.4 and 2% respectively.

27.   World oil prices (in real terms) are used as a proxy for the evolution of
international energy price levels. World oil prices are assumed (moderate
growth scenario) to increase between 1987 and 2000 from $US 18 per barrel to
$US 25-$US 30, and between 2000 and 2010 to a range of about $US 32.50-$US 40.
In the low growth scenario, the increase would be from $US 18 per barrel
in 1987 to a range of $US 20-$US 25 by 2000 and $US 25-$US 30 by 2010.

28.   The various assumptions are related to each other, whereby economic
growth determines energy demand growth consistent with a specified increase of
real oil prices: the increase is higher in the moderate growth scenario than
in the low growth scenario. Oil prices are assumed to determine supply
structures at the margin, whereby the increase of oil prices relative to other
fuels is considered higher in the moderate growth scenario than in the low
growth scenario. These interactions are judgemental rather than formalized in
a model.

C.   DATA BASE

29.   The projections of the present study constitute an adaptation of the most
recent world energy projections prepared by a study group of the World Energy
Council (WEC). 4/ The WEC projections cover 2000-2020 which required an
interpolation of numbers so as to accommodate the time-horizon of the present
study (2010). Also, the country groupings of the WEC did not fully coincide
with the groupings used in ECE. On the other hand, with the exception of data
on non-commercial energy, the WEC relied heavily on UN and ECE demographic,
macro-economic 5/ and energy data. 6/ In choosing the WEC study as the data

base for the present study, its projections are consistent with projections
for other world regions particularly the developing countries, and could be
extended to 2020 if desired.

30. The energy data and balances of the present study refer to production,
exports, imports and apparent consumption at the primary stage only. An
analysis of trends in and implications of secondary energy use, particularly
electricity penetration, is given in Addendum 3 of the present study.

D. DESCRIPTION OF ANTICIPATED ENERGY DEVELOPMENTS

(a) Primary energy demand growth

31. Under the impact of demographic and economic growth, primary energy
demand would continue to rise to 2010, despite a significant decline of the
energy intensities by one quarter to one third (see table 2).

   (i) Growth rates

32. During 1985-2000, the average annual rates of growth of primary energy
demand in the ECE region as a whole would be at 1.0-1.4%. This is above the
average annual growth rate experienced during 1980-1985 (0.7%), but below the
rate recorded in the decade ending in 1985 (1.6%). Growth rates in the market
economies would be 0.7 to 1.2% per annum, in the east European economies 1.4
to 1.8%. During 2000-2010, growth rates of primary energy demand are
projected to decline to 0.3-0.7% per annum (table 2). Under low growth
assumptions, rates in the market economies would practically stabilize at the
level reached in 2000; under moderate growth assumptions, demand would rise by
0.4% on average annually. In the east European economies, growth rates would
decline to 0.8-1.2% during 2000-2010, i.e. by about 50% compared with
1985-2000.

   (ii) Absolute growth

33. Whereas demand in absolute terms in 1985 stood at 5,117 million toe
in the ECE region as a whole, in 2010, demand would be in the range of
6,120 million toe (low-growth scenario) to 6,810 million toe (moderate-growth
scenario). This is an increase of approximately 1,000-1,700 million toe or by
20-33%. By contrast, the increase would be as high as 3,220-4,980 million toe
if the energy intensity of the ECE economies stagnated at the 1985 level.
Improved energy economy and efficiency in the ECE countries is thus projected
to save 2,215-3,290 million toe, according to growth scenario (low to
moderate) (table 2).

   (iii) Regional pattern

34. In 2010, 61% of the primary energy demand of the ECE countries would
originate in the market economies compared with 65% in 1985, due to the
relatively high economic growth rates envisaged for the east European
economies. In 2010, the ECE region would account for 60% of world primary
energy demands, compared with 68% in 1985, - a decrease which reflects the
growing energy demands of developing countries.

ECE REGION
PRIMARY ENERGY DEMAND
million toe

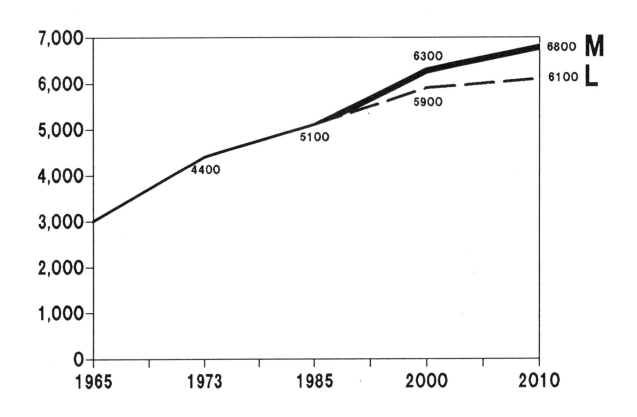

% PER YEAR

AREA	1965-1973	1973-1985	1985-2000	2000-2010
ECE	4.8	1.2	1.0-1.4	0.3-0.7
ME	4.8	0.7	0.7-1.2	0.0-0.4
CPE	5.1	3.2	1.4-1.8	0.8-1.2

(iv) Per capita consumption

35. In 1985, per capita primary energy consumption in the ECE countries stood at 4.72 toe. Up to 2010, per capita consumption would rise, particularly in the east European economies where higher economic and energy growth rates more than compensate for projected population growth (table 2).

(v) Rational use, conservation and demand management

36. The projections assume a large improvement in "energy efficiency" between 1985 and the year 2010 both in absolute terms and relative to past achievements. This improvement is measured in terms of "energy intensity" or energy consumption in tons of oil equivalent (toe) per $1,000 of Gross Domestic Product (GDP) or Net Material Product (NMP). While this indicator measures the energy efficiency of an economy only in the broadest sense, it also reflects many other important features of an economy such as the structure of production, relative share of energy intensive industries, physical activity levels and the actual efficiency of energy production, distribution and end-use.

37. In the past (1973-1985), the energy intensity of industrial economies had fallen by 15%.

38. By the year 2010, reductions in the energy intensity, considered as being viable, could achieve energy savings of 3,290 Mtoe in the moderate scenario and 2,215 Mtoe in the low scenario for the ECE region as a whole. This amounts to a reduction of energy demand by about one third below what it would have been without energy efficiency improvements in both market and east European countries. Demand continues to rise but at much lower rates than in the past due to a continued shift of industrial economies towards service industries, less energy intensive industrial sectors in the structure of production and vigorous energy conservation policies.

39. Greater energy economy and efficiency is anticipated from moderate growth than from low growth. Comparatively high economic growth rates would provide better conditions for rationalizing energy systems and investing in conservation measures. Market economies are likely to be less energy intensive than the east European economies. Greater east-west exchanges in energy efficient technologies are anticipated to contribute to energy savings among all ECE member States.

(b) Changing patterns of primary energy consumption

40. Without exception, all primary energy sources in use in 1985 would be needed to cover projected primary energy demand growth until 2010 (table 3).

41. All primary energy sources are projected to increase in absolute terms until 2010. In the ECE region, and according to the growth scenario, solid fuel use is projected to grow by 260-490 million toe, nuclear power by 280-400 million toe, gas by 300-440 million toe and new and renewable sources by 110-180 million toe. Oil supplies are projected to increase by 50-175 million toe.

42. In the market economies, most of the increments would be coal and nuclear power, in the east European economies: gas, coal and nuclear power.

IMPLIED SAVINGS OF PRIMARY ENERGY IN 2010, COMPARED WITH 1985, ECE REGION

million toe.

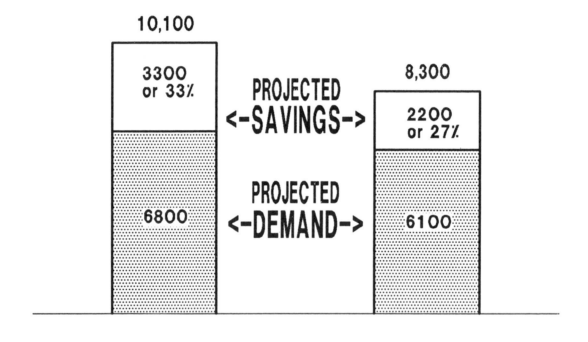

TOE/1000 $/GDP

AREA	1973	1985	2010
ECE	0.67	0.57	0.40-0.43
ME	0.59	0.47	0.33-0.34
CPE	1.03	0.92	0.60-0.71

ECE REGION:
SHARES OF PRIMARY ENERGY SUPPLIES

1985
ECE

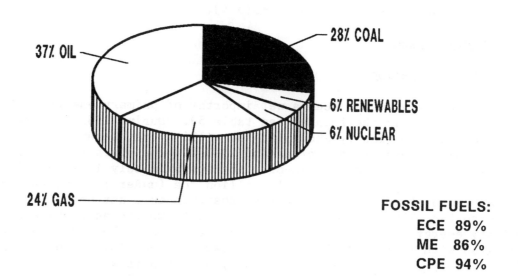

37% OIL

28% COAL

6% RENEWABLES

6% NUCLEAR

24% GAS

FOSSIL FUELS:
ECE 89%
ME 86%
CPE 94%

2010
ECE

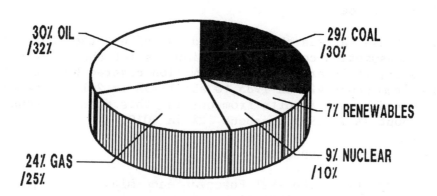

30% OIL /32%

29% COAL /30%

7% RENEWABLES

9% NUCLEAR /10%

24% GAS /25%

FOSSIL FUELS:
ECE 83-85%
ME 79-80%
CPE 90-91%

43.   As a result of differing supply prospects, the <u>market shares</u> of the
various fuels would change (table 4). For the ECE region as a whole, the
share of fossil fuels would fall from 89% in 1985 to 83-84% in 2010, with coal
and gas maintaining or slightly improving their shares, oil losing, and
nuclear power and new sources improving their relative position. The main
difference between the market economies and the east European economies would
be a greater share of fossil fuels in the primary energy balances of the east
European economies (1985: 94.4%, 2010: 90-91%) than in the market economies
(85.8%; 79-80%); gas is projected to further significantly increase its market
share in the east European economies (1985: 30.4%, 2010: 36-36.4%), but to
decline in the market economies (1985: 20.2%, 2010: 17.1-17.3%); in contrast,
the relative role of coal would diminish in eastern Europe (1985: 34.5%,
2010: 30.2-30.8%) but increase in the market economies (1985: 25.1%;
2010: 26.4-27%). In east Europe the market share of new and renewable
sources would rise about 3.5% (1985: about 3%), - less than half the market
shares in the market economies (table 4).

(c) <u>Energy trade</u>

(i) <u>ECE region</u>

44.   The ECE region has been a net importer of primary energy in the past and
would remain so in the future (see table 5). During 1985-2010 net imports
are projected to rise by about 145-150 million toe, depending on growth
assumptions. This is an increase of 29-30%. The overall increase results
from an increase of net oil imports (by approximately 170-215 million toe), an
increase of net gas imports by 12 million toe (under low growth conditions);
and a change of the region's current position as a net coal importer into a
net exporter. The net import dependency (net imports as a share of total
primary energy consumption) of the ECE region as a whole in the year 2010
compared with 1985 would be about the same - about 10% - (in both scenarios).
The share of oil consumption covered by oil imports would rise to about 34%,
compared with 25% in 1985, implying a significant increase of revenue transfer
from oil importing ECE countries to oil exporting non-ECE countries.

(ii) <u>Market economies</u>

45.   Between 1985 and 2010, total net imports of the market economies are
projected to increase by 14-22% due to rising net oil (100-150 million toe)
and gas imports (about 28-35 million toe). The trade deficit in coal
in 1985 (about 30 million toe) would turn into a small trade surplus
(2010: 6-20 million toe).

46.   On balance these various changes would not affect the share of total
primary energy consumption covered by net imports (of about 19% recorded
in 1985). However, the share of oil consumption covered by net oil imports
would increase from about 42% in 1985 to about 49% in 2010. The share of gas
consumption covered by net imports from outside this group of countries would
increase from about 7% in 1985 to about 12% in 2010.

(iii) <u>East European economies</u>

47.   Net energy exports of the east European economies are projected to
substantially decline between 1985 (140 million toe) and 2010 (about
85-130 million toe). This group of countries, which exported 84 million tons

PRIMARY ENERGY TRADE
1985-2010
million toe.

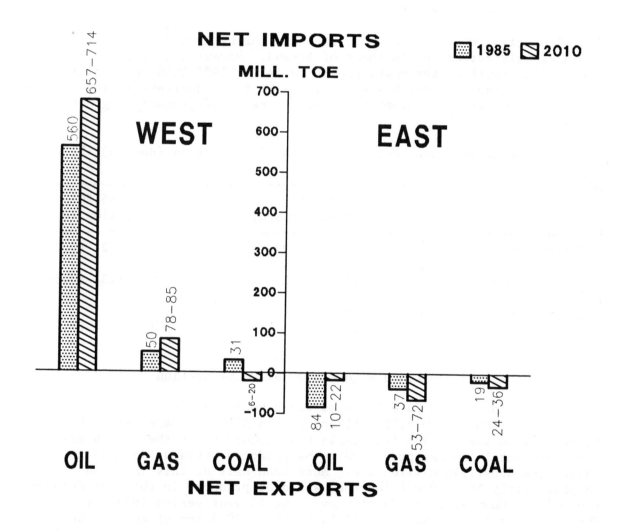

of oil and oil products in 1985, is expected to only export about
10-22 million toe in 2010. However, gas exports would increase from
37 million toe in 1985 to 53-72 million toe in 2010.

(iv) East-west energy trade

48. Because of the vicinity of the east European energy exporters to the
west European energy markets, it can be assumed that the projected changes of
the east European energy exports are indicative of the future size and pattern
of east-west energy trade. The projected decline of exports in energy
commodities underlines the necessity for more diversified energy trade
relations.

(d) Energy production growth and focus

(i) Production growth

49. Production would continue to increase during 1985-2010. Growth rates
would closely correspond to those of demand. However, in comparison with
production growth in the past, growth rates for 1985-2010 would slow down by
half in the moderate growth scenario, and by three quarters in the low growth
scenario. In the east European economies, rates of production growth are
projected to be higher.

Average annual rates of growth of production (demand), %

	1970-1985 production (demand)	1985-2010 production (demand)
ECE region	2.4 (1.9)	0.7-1.2 (0.7-1.1)
Market economies	1.7 (1.2)	0.4-1.0 (0.5-0.9)
East European economies	3.5 (3.4)	1.0-1.4 (1.2-1.5)

(ii) Absolute, incremental and cumulative production

50. Production in absolute tonnages would be impressive: in the ECE region
as a whole: some 5,500-6,200 million toe in 2010, compared with about
4,600 million toe in 1985 (see table 6). According to the growth scenario,
the incremental tonnage projected to be produced between 1985 and 2010 would
range from 860-1,540 million toe. More than half of this increase
(approximately 550-830 million toe) would be produced in the east European
economies. Cumulative production over the 25 year period 1985-2010 in the
ECE region as a whole would be 41-44 billion (10^9) toe of coal, about
35-36 billion toe of oil, and 36-38 billion toe of natural gas (the first
number referring to the low, the second to the moderate growth scenario).

ECE REGION INCREMENTAL PRODUCTION 1985-2010

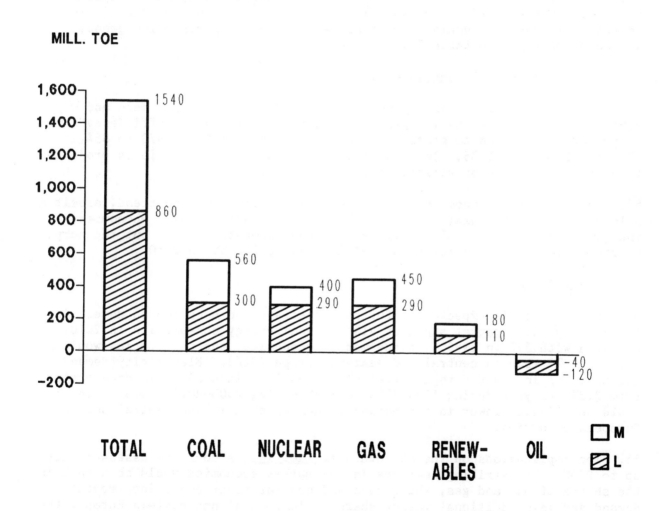

MILL. TOE

(iii) Production pattern

51. The various primary energy sources are not projected to contribute uniformly to production growth; there would even be decreases. Moreover, the growth prospects of the various energy sources are quite different in the market economies compared with the east European economies. In the market economies, most of the production growth between 1985 and 2010 of 315-715 million toe would be accounted for by coal (+190-350 million toe) and nuclear power (+200-275 million toe), followed by renewable sources, including hydropower (+85-140 million toe). Indigenous oil production is projected to fall (-50 to -100 million toe or by -6 to -13%). Indigenous gas production would stabilize under moderate growth conditions, but fall under low growth assumptions (-55 million toe, or -8%) (see table 6). In the east European economies, most of the additional production of 550-825 million toe by 2010 will be in natural gas (+340-435 million toe), followed by coal (+115-210 million toe), nuclear power (+90-130 million toe) and renewable sources, including hydro (+23-43 million toe). Oil production has already peaked. Gas and coal would surpass oil as the most important indigenous source of energy (see table 7).

(iv) Role of "self-sufficiency"

52. Most of the primary energy needs of the ECE countries has been and would continue to be covered by indigenous production. The "self-sufficiency" ratio of primary production to gross consumption rose from 83% in 1970 to 85% in 1980 and to 90% in 1985. In 2010, this "self-sufficiency" ratio is projected to be about the same or slightly less as in 1985.

53. In the market economies, the ratio would follow the same trend, albeit at a lower level (1985: 81%; 2010: 80-81%). In the east European economies, indigenous production would continue to exceed domestic needs, but the surplus of production would decline from 7.9% in 1985 to 3.6-5% in 2010.

(e) Electrification

54. Electricity is expected to further penetrate end uses and to cover, in the ECE region, a market share of 41% of final energy consumption in 2010, compared with 32% in 1985. In the market economies, the penetration would be higher (48%) than in central and eastern Europe (32%). Electricity demand growth rates in the ECE region (calculated as fuel inputs) would decelerate from 2.3% per year during 1985-2000 to 1.8% during 2000-2020. Growth rates would be slightly lower in the market economies than in the central and east European countries.

55. The implications on the structure of fuel inputs could be assessed only up to 2000. The striking features in the market economies would be a fall in the shares of oil and gas, while coal and nuclear power cover incremental demand and gain additional market shares. In central and eastern Europe, the shares of oil, coal and gas are expected to fall (with gas and coal increasing supplies in absolute terms, however); the underlying assumption is that nuclear power would increase its relative contribution to electricity generation in central and eastern Europe from 10% in 1985 to 25% in 2000.

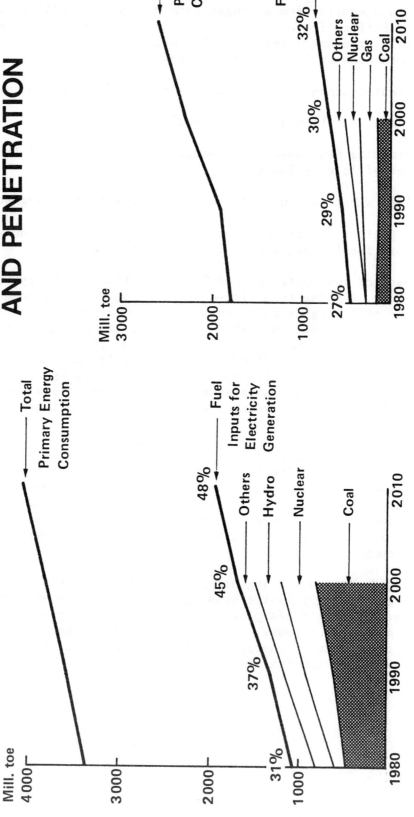

ELECTRICITY DEMAND AND PENETRATION

E.   THE IMPACT OF ENVIRONMENTAL PROBLEMS ON ENERGY PROJECTIONS

56.   The above energy projections reflect prevailing environmental concerns and policies in ECE member countries by the end of the 1980s. These concerns concentrated on SO_2 and NO_x emissions, water pollution, dust and noise. The energy implications of remedial action with regard to these pollutants are "built into" the national energy projections, either on the basis of actual regulations or of anticipated measures. Whether the environmental concerns "captured" are adequate from a longer term perspective will be considered in Chapter III. Suffice it to say at this stage that the containment of emissions of heavy metals, radionuclides, volatile organic compounds or CO_2 had not been taken into account in the above energy projections.

57.   Environmental concerns affect energy projections in four ways:

- by enhancing the need for rational use of energy;

- by creating new demand for energy to operate control techniques or to offset efficiency losses;

- by fuel substitution in favour of the least polluting sources;

- by impacting on the location of energy production, conversion or end-use facilities, and on international trade.

58.   The impact of environmental concerns on efficient energy use appears to be most striking, leading to a slow-down of primary energy demand growth. Savings exceed the energy requirements of the environmental control techniques involved (catalysts, desulphurization/denitrification facilities). The latter energy requirements are indeed low: for example, in the case of flue gas desulphurization, 0.75% to 2.7% of gross electricity production in the United States of America and 1-1.5% of gross fuel consumption in the Federal Republic of Germany. 7/ In comparison with these additional energy requirements, the losses in system's efficiency and plant availability may be much more important.

59.   The impact of sulphur emission standards on fuel substitution - certainly significant at the enterprise level - appear less evident at the aggregate levels of the above energy projections. The ECE region and subregions (though not all countries) are well endowed with low sulphur coal, oil and gas and, in case of need, dispose of effective abatement technologies (flue gas desulphurization; fluidized-bed combustion; integrated coal gasification/combined cycles; selective catalytic reduction).

60.   Energy producing/converting/using facilities are sensitive to land-use constraints, particularly in the densely populated areas of west Europe and North America. Siting issues place constraints on the growth prospects of hydro, solar and nuclear power. There is no evidence that these constraints motivate the relocation of energy producer/user activities towards other countries with lesser constraints.

CHAPTER III

APPRAISAL OF THE SUSTAINABILITY OF THESE PROJECTIONS AND SCENARIOS

A. THE APPROACH

61. Are the projected energy paths "feasible", "viable", "sustainable" during the period to 2010 and from a longer term perspective?

62. The question poses a number of conceptual problems: how to define "sustainability", "viability", etc. in an international, particularly east-west context; how to avoid a bias when selecting certain criteria rather than others; how to relate the various criteria to each other, hierarchically or otherwise?

63. Whatever approach is chosen, it will partly prejudge the results of the appraisal.

64. In order to mitigate this risk, the present chapter, following the pragmatic approach adopted by the Preparatory Meeting (ENERGY/AC.10/2) applies a large number of criteria of different nature. After a review one by one (Section B), a synthesis (Section C) is given to provide a listing and hierarchy of constraints associated with the projected energy paths.

B. CRITERIA APPLIED

(a) Capacity to meet aniticipated demand

65. The overriding attraction of the two energy scenarios for the ECE region up to 2010 resides in the message that the quest for energy services of a rising and increasingly affluent population could be met within the limits of the time horizon and assumptions made. Energy need not be a constraint.

(b) Security of supplies

66. The security of energy supplies would remain high. However, while the net import dependence (all fuels taken together) of the market economies would remain unchanged at about 19%, the share of oil consumption covered by oil imports would rise from 42% in 1985 to 49% in 2010. A related concern is whether the incremental oil imports of 100-150 million toe would originate from several suppliers or from one region. These concerns will be further exacerbated by the expectation that the oil imports of the market economies from the east European economies would fall from 84 million toe in 1985 to about 10-22 million toe in 2010, depending on growth assumptions.

(c) Resilience against temporary, localized supply disruptions

67. The resilience of the ECE energy economies against temporary or local disturbances, due to weather conditions, unplanned outages of plants, etc., remains high, but appears to weaken somewhat.

68. Excess supply capacities available in the market economies throughout the 1980s will have been used up in the 1990s; land use constraints and public resistance are likely to slow capacity growth. In the east European economies, capacities would continue to be under strain throughout the 1990s, primarily from lack of capital but growingly also as a result of public opposition. Stocks of oil, coal and gas, ample in the market economies, are likely to remain short in the east European economies.

69. On the other hand, the infrastructure for international trade in energy and international exchanges of electricity would strengthen, probably even to a greater extent than the domestic infrastructure. This enhances the resilience of the co-operating systems in the face of local or temporary disruptions. The importance of this interdependence should not be overestimated, however. Throughout the period under review, the primary and secondary energy systems of the ECE region would remain essentially "national" systems. Strengthened trade links cannot make up for delays in domestic capacity growth.

(d) Compatibility with energy reserves, resources

70. World primary energy "reserves", defined as "proven recoverable amounts in place that could be recovered under present and expected local economic conditions with existing technologies" (World Energy Council: 1986 Survey of Energy Resources, p.11) would suffice to meet projected world and ECE demand by 2010 (table 8). Significant reserves and resources would remain for utilization in subsequent centuries (coal) or decades (gas, uranium). By 2010, only about 20% of the world's hydropower potential and about 50% of the hydropower of the ECE region would have been harnessed.

71. Conventional oil reserves, however, are likely to come under strain in the new century, if not at the world level, certainly in the ECE region where cumulative demand during 1985-2010 would largely exceed indigenous reserves. These pressures will lead to a growing exploitation of available conventional and unconventional resources (heavy oil, tar sands, oil shale) or technologies (enhanced recovery, deep offshore and Arctic exploitation). Under the price assumptions of the present study, some of these resources would become economically viable by the time they are needed.

(e) Implied progress in energy economy and efficiency

72. Reductions in the energy intensity of industrialized economies by 25 to 30% by the year 2010, as projected in the present report, appear to be feasible and viable projections provided that public awareness, national and international energy efficiency programmes are maintained at or above levels achieved between 1973 and 1985. The level and rate of reducing the energy intensity (toe/$1,000 of GDP in 1980 dollars) of industrial economies for the recent past and the future is shown below:

	1973	1985	2000 M	2000 L	2010 M	2010 L
Market Economies	.59	.47	.38	.39	.33	.34
East European Economies	1.03	.92	.71	.80	.60	.71
Total ECE Region	.67	.57	.46	.48	.40	.43
Index 1985 = 100	118	100	80	84	70	75

73. Some might consider the projected reduction as conservative given the findings of recent international studies. One report based on case studies of Brazil, Sweden, India and the United States shows that energy consumption per capita in industrial countries could be reduced by 50% by the year 2020 while GDP per capita grows by 50 to 100%. 8/ In contrast, energy consumption per capita rises by 4 to 15% between 1985 and the year 2010 in the present study, with per capita GDP rising by between 60 and more than 100%. This is not to suggest that the present study denies the possibility of a low energy future for industrialized countries - it is technically possible - but it contends, on the basis of past experience, that the inertia of systems and behaviours affect the level and rate at which energy efficient technology and demand management practices can be introduced. Clearly, greater exchanges among industrial countries would help to ensure the rapid and widespread dissemination of the many proven energy efficient technologies which are commercially available throughout Europe and North America today. Government policies could assist this process, e.g. by means of incentives and internationally-agreed efficiency standards.

74. This would be particularly true in the east-west context. As can be seen from the above table, progress in reducing the energy intensity of the economies is comparable between the market economies and the east European economies. However, an efficiency "gap" would persist. One might have expected a greater reduction of the energy intensities of the east European economies, given the higher potential for improvements. If the scenarios do not show such a tendency, it is not only due to the high economic growth rates in the East and to a more energy-intensive growth path, but also to the prudent assessment of future energy technology transfer from the market economies to the east European economies.

(f) Viability, cost effectiveness of resource allocation

75. Are the scenarios economically viable? Are resources allocated in a cost effective manner? Are the projected roles of the various fuels, of energy supplies versus savings, of energy use versus the use of other production factors (labour, capital or "the environment") economical?

76. If related to the prevailing (deficient) pricing and costing systems, the reply is certainly affirmative, on account of the dominating role assigned to market forces, in particular oil prices, in operating the projected demand/supply adjustments.

77. The evaluation is different if the deficiencies of the present pricing and costing systems, not only in the east European economies but also in the market economies are taken into account. These deficiencies consist primarily of a deviation of the actual energy prices from long-term marginal replacement costs, of the exclusion of the so-called social costs or externalities from prices, and of impediments to market access. */ These deficiencies affect the scope and structure of the energy projections to an extent which cannot be quantified but appears notable. Accounting for the deficiencies would have placed an onus on energy supplies (in favour of energy savings), on primary energy use (in favour of secondary energy sources) and among the various supply options on the most subsidized and protected sources of energy (in favour of the others). Accounting for the deficiencies would have accelerated the rise of energy, particularly oil prices, and would have affected the projected energy path in the east European economies to a greater extent than in the market economies.

(g) Rising energy supply costs

78. The projected energy developments may be "viable" and "cost-effective" — they nevertheless imply an increase of the cost of energy to the end consumer. The increase is determined by the projected need to provide capacities in response to rising demands and by the projected rise of specific investment costs resulting particularly from a worsening of the resource base and the application of environmental control techniques. Also, investments in the rational use of energy or interconnections, while they may be more cost-effective than investments in incremental supply capacities, nevertheless increase front-end investments.

79. Total cumulative investments of the ECE energy supply industries in production, conversion, transportation and distribution of energy during 1980-2000 are estimated to be between $US 7,700 billion and $US 8,900 billion, of which more than two thirds for electricity generation. This corresponds to an overall growth of investment needs of 22-40% for the ECE region — less for the market economies (+16-34%), more for the east European economies (+39-97%). Annual growth rates would be 1-2% for the ECE region, depending on growth scenarios, 0.8-1.6% for the market economies and 2-5% for the east European economies. These growth rates are below the projected growth of GDP, but within the range of NMP projections. This suggests that the investment "drain" by the energy supply industries tends to ease in the market economies (but not in the east European economies), and that the rise of energy costs for the energy consumer, while significant in absolute terms, would not be disruptive in relative terms, at least not in the market economies.

---

*/ Solely with regard to the "Single Market", the Commission of the European Communities identified 56 "obstacles" (COM (88) 238 final, 2 May 1988).

80. Among the various factors causing the projected increase of investments and of running costs, environmental controls are likely to play the determining role. They will be necessary at all stages of the energy economy: exploration, extraction, conversion, transportation and use and waste disposal. Costs involved are not known for entire energy systems, but only for certain segments. For example, capital costs of environmental control techniques for SO_2, NO_x, particulates, volatile organic compounds, liquid effluents, noise, solid residues and for landscape rehabilitation are estimated to increase the total investments for new coal-fired power stations by 30-35% of which 15 and 6% respectively for flue-gas desulphurization and selective catalytic reduction. Retro-fitting of existing coal-fired plants with flue-gas desulphurization would add 10-40% to capital costs; selective catalytic reduction appears less expensive. 9/ Environmental controls would add 15% to total investments in a modern coking plant with a throughput of 2 million tons ("Prosper" in Bottrop, FRG).

(h) Technical feasibility

81. The projected scenarios are technically feasible. Proven technologies enhance the recovery of oil in place; desulphurization technologies capture up to 90% and more of the sulphur content of coal; efficient building design saves 30-40% of energy compared with pre-1980 dwellings; computers optimize the operation of energy appliances and facilities; engines under development would reduce average gasoline consumption per km by one quarter to one third by 2010.

82. The problem would seem to lie in an adequate technological preparedness for the longer term, beyond 2010-2020. The technologies of the twenty-first century must be developed well in advance, not only to ascertain the technical maturity and economic viability, but also to allow for the lengthy process of application and market penetration. It is not evident that these conditions are being met with regard to technologies thought to be needed in the longer term: fusion, breeders, seasonal storage of cold and heat, CO_2 removal and disposal techniques, large-scale hydrogen systems, deep drilling and long distance transportation of gas, the direct conversion of methane to methanol, photovoltaics. Worse: international co-operation in energy R & D has weakened in comparison with the late 1970s/early 1980s.

(i) Implications of projected energy trade developments on general trade

83. Energy is one of the most important commodities of world trade, accounting for some 18% of world trade in 1985. The projected increase of energy imports by the market economies, as well as the projected reduction of net energy exports of the east European economies must, therefore, be assessed for their implications on general trade relations.

84. During 1985-2010, the market economies are projected to increase net oil imports by about 100-150 million toe, to increase net gas imports by about 28-35 million toe and to transform the coal trade deficit of about 30 million toe in a surplus of 6-20 million toe (table 5).

85. In 1985, the net oil imports of the market economies of 560 million toe provided revenues to oil exporters and traders from third countries of the order of $US 113 billion (based on an average cif-ARA crude oil posting of $US 27.59/barrel for 1985). In 2010, the above-mentioned increase of net oil

imports would create revenues for oil exporters and traders under the price assumptions of the present study (para. 27) of between $US 184-221 billion (low growth scenario) and $US 260-320 billion (moderate growth scenario). The projected increase of gas imports would provide revenues of between $US 48 (low growth scenario) and $US 62 (moderate growth scenario) compared with $US 19 billion in 1985. In a nutshell, moderate growth conditions imply an increase of capital transfer from the market economies to the oil exporting countries of between $US 25-$US 105 billion. The gas import bill of the market economies would remain practically unchanged under moderate growth conditions and even fall under low growth conditions.

86. The projected overall growth of the energy trade between the market economies and third countries would provide a continued stimulus for an expansion of general trade.

87. The net energy exports of the east European economies are projected to decline between 1985 and 2010 by 6% (moderate growth scenario) and 39% (low growth scenario). Increased gas and coal exports would not make up for the significant reduction of oil exports. All other conditions unchanged, a loss of revenues would ensue which, seen the important role of energy exports as hard currency earners for the east European economies, could not but affect their capacity to finance imports. Once more, the diversification of exports of this group of countries, away from basic products and energy, emerges as a requirement of paramount importance.

(j) Associated health, environmental and climatic hazards

88. Primary energy extraction and supply are a precondition for the development and welfare of the ever-increasing population of the industrialized countries. As any other industrial activity, it carries risks for human health and ecosystems. In this regard, the period under review (1985-2010) does not differ qualitatively from earlier periods. It does differ, though, in quantitative terms: on account of the intensification of energy use, the risks would tend to become growingly international and inter-generational, unless preventive measures were taken.

89. The extent and nature of energy-related health and ecological risks depends on the energy strategy pursued. Different energy strategies imply different risks. The scenarios retained in the present study comprise the following hazards:

(i) Hazards associated with technological energy conservation

90. Energy economy and efficiency in the ECE countries is projected to increase by one quarter to one third over the period 1985-2010, due to technological improvements, structural changes of the economies and energy-conscious consumer attitudes and Government policies.

91. Savings induced by new or improved technologies imply health and environmental risks associated with the extraction, upgrading, transformation and use of the necessary raw materials. These risks are certainly on the increase. An example are heat pumps whose pumping media need to be environmentally safe.

(ii) Hazards associated with energy production growth

92. The ECE countries are projected to increase energy production till 2010 by about 860-1,540 million toe, depending on growth assumptions - an increase of 19-33% over 1985 (see table 6). Unless controlled, the exposure to health and environmental risks would grow at much higher rates due to the projected focus on production rather than on trade (and production in ecologically less sensitive regions), the deterioration of geological conditions and the introduction of more sophisticated or new exploitation techniques.

93. To point to a few likely developments and implications: underground mines would reach depths never attained before, requiring the control of higher temperatures, greater mine-water intakes and rock pressures. The volume of overburden and the depth of opencast mines would increase significantly enhancing the disruption of landscape, biotopes, flora and fauna and of ground water and surface horizons. Underground gasification if introduced in the period under review on a commercial scale is likely to lead to leakages of combustion gases and uncontrolled ignition. The number of oil and gas rigs operating in a sensitive natural environment, in deep waters or in the Arctic, would steeply rise as would related risks. Enhanced recovery techniques employing chemicals would become a matter of routine, as would the need for containment. Geothermal exploitation would extend into greater depths and would tap sources with a greater content of noxious compounds. The large-scale use of hydrogen by a broader public presupposes the solution of safety risks. The broad use of solar and wind facilities requires considerable space requirements. Oil-shale mining and retorting, a growth industry, would growingly affect land, water, air, biotic systems and health, if uncontrolled. The scope and magnitude of waste and by-products resulting from energy production would rise over-proportionally and enhance the issue of the safe disposal or retreatment of residues.

(iii) Hazards associated with the fuel pattern of consumption

Fossil fuels

94. The primary energy supply pattern of the ECE countries is projected to remain dominated by fossil fuels. Consumption of coal, oil and gas at 4.5 billion toe in 1985, is projected to grow to 5.2-5.7 billion toe in 2010, depending on growth assumptions (see table 3). This is an increase of 13-24%. The share of these fossil fuels is projected to fall, however, from about 89% in 1985 to 83-84% in 2010. The relative decrease of fossil fuels is projected to be more noticeable in the market economies than in the east European economies and slightly more pronounced under moderate growth conditions than under low growth conditions (table 4).

95. The risks involved by the projected 13-24% increase of fossil fuel use are those of generation of dust, sulphur dioxide, nitrogen oxide, heavy metals, hydrocarbons, carbon dioxide, overburden and waste, including waste from environmental control facilities. The impact of the risks depends on the type of fuel used (natural gas is associated with lesser pollution than coal and oil), the degree of concentration of supply activities (a multitude of small producers and users is likely to produce more nuisances than a few, bigger facilities), the depth of transformation (gasification, electrification provides better scope for control than primary energy use), as well as on emission standards, control technology and availability of finance.

96. With <u>regard to the retention of dust, sulphur dioxide and nitrogen oxides</u>, effective control techniques and regulatory standards, including international, 10/ are projected to be applied generally, thereby lifting the constraints that otherwise might limit the projected growth of fossil fuel use. It is assumed that existing standards would be strengthened and that in addition emission standards for compounds not yet covered such as diesel soot particles, heavy metals and hydrocarbons, and deposition standards for toxic wastes, would be internationally set. The main impediment does not appear to be technical, but economic since the control techniques to be employed increase production and conversion costs significantly (see section (g) above).

97. For a number of years an extensive discussion has centred around the so-called "<u>Greenhouse effect</u>". The concentration of various gases, including CO_2 from agricultural activity, deforestation and fossil fuel use, is seen to increase global mean temperatures, provoke climate variation, and raise the sea level thus affecting the living conditions of future generations. One element in this complex equation of world climate and development is the use of fossil fuels, particularly solid fuels which release more CO_2 per unit of heat than oil and natural gas. Whether the projected growth and pattern of fossil fuel use in the industrialized countries would be sustainable from a climatic point of view, is a matter under consideration and research, particularly regarding the absorptive capacity of CO_2 sinks, and the relative impact of fossil fuel use on total greenhouse gas emissions. A marginal airborne fraction model applied to the projections of the present study warns that a doubling of pre-industrial CO_2-concentrations would occur before the end of the twenty-first century, and much earlier if other greenhouse gases and non-fossil fuel carbon sources were taken into account. Under these circumstances, it is suggested that the risk of a significant temperature increase is such that preventive measures need to be contemplated before the cause-effect mechanisms are clearly established and quantified.

Hydropower

98. Hydropower is projected to be constrained by the growing shortage of sites, environmental questions, alterations of hydrological régimes and by high capital costs. The fact that hydropower is pollution-free may explain the steady long-term growth of hydropower in the ECE countries, from the equivalent of 287 million toe in 1985 to 370-402 million toe in 2010. Its share in total primary energy consumption would increase from 5.6% to 6%.

Nuclear power

99. The ECE countries are projected to increase nuclear power use from 283 million toe in 1985 to 567-687 million toe in 2010, depending on growth assumptions (see table 3). The share of nuclear power in the primary energy balance of the ECE countries would almost double, from 5.5% in 1985 to 9.3-10.1% in 2010.

100. The health and environmental hazards associated with nuclear power are of different nature, depending on the type of operation, the technology employed, the control techniques, the climate, etc. Risks associated with the (underground or opencast) mining and milling of uranium ores are comparable to those of fossil fuels, except for the possible exposure to radon daughters during mining and radiation from inactive mill tailings piles. Uranium conversion, enrichment and fuel fabrication do not have major environmental

impacts. Routine low level or accidental release of radiation are the main
risks during reactor operation. The process of power generation results
in heat release and, hence, thermal pollution. Reprocessing reduces the
occupational and environmental risks of front-end operation while the radiation
generated during reprocessing is low compared with levels currently accepted;
however, the generation and recycling of plutonium during reprocessing
requires additional safeguards against theft and misuse. Finally at the back
end, high level wastes and spent fuels have sufficiently persistent biological
hazards to require special long-term isolation.

101. Among these hazards, reactor accidents and the long-term storage of
high-level wastes and spent fuels are perceived as the main risks. The
projected growth of nuclear power would seem to depend in large measure on the
strict application of the highest available standards of nuclear safety in all
its aspects particularly operation, waste management and new technologies
including new safe reactor technology. Improvements in these areas would
bring to the fore the environmental opportunities that nuclear power offers
thanks to the absence of sulphur dioxide, nitrogen oxide and carbon dioxide
emissions. Particularly the latter advantage might weigh in public opinion
and policy considerations, as and when the implications of climate variation
are more generally perceived as grave and lasting.

(k) Compliance with generally-shared policy objectives, societal and economic
    trends, and public concerns

102. The scenarios of the present study have not been developed in isolation
from government policies, public conerns and societal and economic trends.

103. They emphasize the same priorities as ECE Governments: the need to meet
energy demand, the preference given to enhanced energy economy and efficiency,
the shift among supply options in favour of new and renewable sources, the
importance of a high degree of self-sufficiency and trade diversification, the
reliance on market forces in operating adjustments, etc.

104. They also comply with ongoing societal and macro-economic trends,
characterized by the emergence of service and information-oriented societies
and economies. Such systems emphasize energy services (rather than
commodities), require the instantaneous and ubiquitous availability of energy,
depend on a high level of quality assurance (security of supplies, narrow
frequency fluctuations) and are integrated into world-wide energy systems of
gigawatt dimension. Service and information-oriented economies favour
electricity as a vector, not only of power but also of information. The
scenarios of the present study take these trends into account: the share of
primary energy which reaches the end-consumer in the form of electricity is
anticipated to grow from 32% in 1985 to 40% in 2010.

105. Lastly, the scenarios are committed to public concerns, not to
forecasting. They are instruments for the early identification of
bottle-necks, inconsistencies and risks. The previous sections have
identified these risks as being: the climatic implications of fossil fuel
use; the safety and waste disposal aspect of nuclear power; the health risks
resulting from pollution, plant failures and day-to-day operations of energy
appliances and facilities; concerns about supply disruptions; competing
policies; inadequate international commitments to energy R & D, transfer of
energy-efficient technologies, or multi- or bilateral financing of
environmental control techniques.

C. OVERALL APPRAISAL

106. In the short- and medium-term till about 2010, the projected energy developments appear feasible, efficient, secure and viable options; their environmental impact could be controlled if the appropriate measures were taken. Drawbacks exist, though, and gain importance. Growingly and if extrapolated further into the twenty-first century, the projected scenarios appear unsustainable, as long as the technological and institutional response to the perceived risks, particularly of climate variation and depletion of presently-known conventional energy reserves, remains inadequate.

107. Within its time horizon to 2010, the projected scenarios are <u>attractive</u> in many regards. They demonstrate that energy strategies could be designed which

1. meet the energy service requirements of a growing and increasingly affluent population;

2. maintain security of energy supplies at a high level;

3. assign a greater role to the rational use of energy than to all supply options taken together;

4. are compatible with energy reserves and resources in the world and, except for oil, the ECE region;

5. provide a growing role for new and renewable sources of energy;

6. operate adjustments cost-effectively, primarily through market mechanisms supplemented, where necessary, by government regulations and incentives;

7. suggest that the rise of energy costs to the end-consumer need not be disruptive;

8. are technically feasible, not only with regard to enhanced energy economy and efficiency and incremental energy supplies, but also with regard to health and environmental controls;

9. support world trade and international division of labour; and

10. comply with generally shared energy policy objectives and societal and economic trends.

108. The projections further assist in the early identification of <u>risks</u>. These include:

1. an increase of the oil import dependence of the market economies;

2. a decrease of the resilience of ECE energy systems against temporary and localized supply disruptions, due in large measure to constrained capacity growth;

3. the persistence of a significant efficiency "gap" between member
    countries, particularly the market economies and the east European
    economies;

4. the depletion of the presently-known "conventional proven oil
    reserves" and "additional oil resources" (World Energy Conference
    categories) in the market economies;

5. a slow and from a longer-term perspective low penetration of new and
    renewable sources into the primary energy balances, particularly in
    the east European economies;

6. imperfections of the market mechanisms, such as obstacles to energy
    trade and co-operation, pricing at short-term rather than long-term
    marginal replacement costs and exclusion of externalities from price
    formation;

7. constrained possibilities for financing capacity growth and
    environmental control techniques in some market economies and in the
    east European economies generally;

8. a contraction of east-west trade volumes in energy commodities
    likely to cause a contraction of general east-west trade (low growth
    scenario);

9. increased exposure to health and environmental risks and economic
    rather than technical impediments to the application of control
    techniques;

10. major uncertainty for fossil fuel use resulting from the perceived
    long-term risk of climate change; and

11. uncertainty about the rate of nuclear power growth which would seem
    to depend on the strict application of highest safety standards
    throughout the nuclear fuel cycle, and on the environmental
    opportunities (absence of SO_2, NO_x, CO_2 emissions) of this
    source of energy.

109. These drawbacks would not jeopardize the genuine sustainability of the
described energy paths till about 2010. However, an extrapolation of these
strategies further into the twenty-first century would imply an increase of
CO_2-concentrations from fossil fuel combustion to levels thought to cause a
significant rise of temperatures and sea-levels. In the long term, an
extrapolation of the projected trends would also lead to the depletion of
presently-known conventional oil, uranium and gas reserves. Both constraints
are so significant that scenarios of the type presented do not appear
"sustainable" in a long-term perspective.

CHAPTER IV

ADAPTATION OF PRESENT ENERGY DEVELOPMENTS AND POLICIES

110. The previous chapter has shown that remedial action is now warranted to cope with certain drawbacks of the projected development up to 2010 and, more importantly, to secure their sustainability well into the twenty-first century. The earlier such remedial action is implemented, the less stringent and costly it will be. Therefore, the focus of the present chapter shifts from projections and perceived risks to the identification of early policy responses. This chapter considers first the various measures in isolation and next, preferred policy mixes.

A.   MARKET RESPONSE AND POLICY CRITERIA

111. As detailed in paragraphs 75-77 above, market forces are assigned the central role in securing the economic viability of the projected energy developments till 2010, despite deficiencies such as impediments to market access, deviation of actual prices from long-term replacement costs and the exclusion of externalities from price formation. While these market imperfections may have been tolerable in times of ample resource availability, they are no longer acceptable from a longer term perspective: as time goes by, they introduce an ever-growing bias in the use of production factors and ecosystems.

112. Elements of remedial action already emerge. Policies and discussions in the market economies focus on the abolition of obstacles to market access (the "Single Energy Market" of the European Community; the US/Canadian Free Trade Area) and on user taxes to incorporate externalities into energy prices. In the east European economies, end-user prices are significantly increased to better reflect production costs and world market prices.

113. Adjustment policies of the kind described prove difficult however. They require the abolition of protectionist measures for certain fuels or industrial areas, the quantification of externalities, a resource-effective assignment of penalities, and above all a new balance between conflicting policies: optimization of resource allocation, also between generations; social equity; international competitiveness of national economies; overriding environmental demands of the international community. Whatever the difficulties: <u>the reduction of market imperfections would appear to be the most effective single systemic measure to adjust present energy consumption and supply trends to long-term scarcities</u>, supplemented, where warranted, by government regulations and incentives.

B.   ENHANCED ENERGY DEMAND MANAGEMENT

114. Second in importance are efforts to increase energy economy and efficiency. The projections already imply a significant reduction of the energy intensity of the ECE countries of about one quarter to one third up to 2010 (table 1).

115. A supplementary effort could consist, for example

- in decoupling energy and economic growth as of 2000, and/or

- in reducing the efficiency "gaps" between member States, particularly the market and east European economies, say by half, by 2010.

The first approach focuses on further technical progress, the second on trade facilitation.

116. Such an additional decrease of the energy intensity is technically possible and could be implemented by technological breakthroughs (such as super-conductors; integrated combined cycle coal gasification and combustion), the general application of known technologies (such as combined heat and power generation), further structural adjustments of the economies, enhanced energy-conscious attitudes of the public at large, traffic planning and more stringent institutional pressures such as the establishment of minimum efficiency standards for mass-produced machinery, vehicles and equipment. These measures would probably not be cost-effective at the assumed rise of oil prices, but presuppose even higher oil prices and/or deliberate government interventions. Whatever the related uncertainties, the impact of such a decrease of the energy intensity on the lifetime of reserves, on investments, air and water pollution and CO_2 emissions would be significant:

- under moderate growth conditions, the uncoupling of energy demand and economic growth in the entire ECE region as of 2000, avoids energy use of as much as 541 million toe or 7% in 2010, of which 83% fossil fuels;

- under moderate growth conditions, a reduction of the efficiency gap for example between the market economies and the east-European economies of eastern Europe by 50%, "saves" 540 million toe in 2000 and 604 million toe in 2010 of which 90% would be fossil fuels; an even greater saving would result from a reduction of the efficiency gap between North America and Western Europe.

117. At the same time, the difficulties and conditions of such additional efforts should not be overlooked. Supporting policies would become necessary, to overcome the inertia of systems and behavioural patterns, comprising possibly:

- energy pricing at long-term replacement and "social" costs;

- the setting of international minimum efficiency standards;

- the granting of government incentives for the development and application of energy-efficient technologies, particularly in times of low oil prices;

- the facilitation of technology transfer, trade, industrial co-operation and international research in energy-efficient technologies;

- multilateral financial assistance for the transfer of energy-efficient technologies and practices.

118. Whatever the difficulties, the anticipated benefits would justify a broad international exploratory effort in this direction.

C.   ENHANCED ENERGY SUPPLY MANAGEMENT

119. The stable supply of energy to the ECE region requires the consideration of the following measures:

    (i)   to maintain exploration for hydrocarbons and uranium in the ECE region at a high level;

    (ii)   to improve the cost-effectiveness of technologies for the extraction and conversion of unconventional, high-cost oil resources and substitutes;

    (iii)   to abolish obstacles to energy trade and co-operation;

    (iv)   to facilitate the multilateral financing of the development of energy resources and growingly, application of energy-efficient technologies;

    (v)   to favour a faster-than-projected market penetration of new and renewable sources of energy, particularly solar energy;

    (vi)   to enhance research in technologies which would assure the large-scale use of new and renewable sources for high energy density purposes or areas, in particular research into seasonal storage of solar energy and solar/hydrogen systems.

D.   ENHANCED ENVIRONMENTAL AND HEALTH MANAGEMENT

120. In this regard, the following measures deserve attention:

    (i)   to pay greater attention to the rising health and environmental risks of routine energy activities, including technological energy conservation;

    (ii)   to enhance research into the climatic implications of various energy supply mixes and into the need for, and possibilities of, measures to reduce, capture, recycle or safely deposit CO_2;

    (iii)   to extend the international harmonization of emission standards to compounds not yet covered and to the deposition of toxic wastes;

    (iv)   to provide multilateral assistance in the financing of environmental control techniques;

    (v)   to promote the international harmonization of nuclear safety standards on the basis of best available, viable technology;

    (vi)   to enhance co-operation in the development of inherently safe reactor technologies, the safe, long-term disposal of high-level radioactive wastes and spent fuels, and the introduction of resource-efficient reactor lines.

E.   PREFERRED POLICY MIXES

121. So far, the various measures have been reviewed one by one, even though in all probability Governments would need to combine various measures in order to secure sustainable strategies. What are the preferred policy mixes, nationally and internationally?

122. While there are likely to be numerous variants of preferred policy mixes, particularly at the national level, the global constraints of climate variation and depletion of conventional energy reserves appear to narrow the choice to "five essentials":

- the rational use of energy, as the most important option;

- a reduction of the relative importance of fossil fuels in favour of non-CO_2-emitting sources, provided their impact on the environment is acceptable;

- among fossil fuels: a relative shift in favour of gas;

- the development of environmentally, particularly CO_2 benign, control technologies.

- a continued assessment of the environmental impacts of nuclear energy.

123. In the present study a methodology has been tested to link the global constraint of climate variation with a global energy response based on the above "five essentials". The value of the exercise lies perhaps more in testing an analytical tool (which could be used to evaluate numerous international response strategies) and in understanding the interplay of the various influencing factors, than in its necessarily tentative conclusion. Still, it is shown quantitatively that an evolutionary adaptation of the present world energy system can buy sufficient time for lasting solutions to be found to the reduction, removal and disposal of CO_2, while admitting an increase of CO_2-concentrations of about 1.75 times the pre-industrial level. The scenario does not assume excessive reliance on solar or nuclear power or an absolute reduction of coal use, but features rising consumption of all energy sources, with preference given to non-fossil fuel sources and within the latter to the least carbon emitting. The message is affirmative: it could be done.

CHAPTER V

THE ROLE OF REGIONAL ENERGY AND ENVIRONMENTAL CO-OPERATION, PARTICULARLY THROUGH ECE

124. Integrated and sustainable energy and environmental policies require strong and increased international co-operation. The constraints to sustainability are not only region-wide but mainly global. National response strategies may be rendered less effective unless internationally consistent. This applies particularly to the many small and medium-sized countries of the ECE region.

125. Global responses to the two overriding threats to long-term sustainability - climate variation and resource depletion - are being developed. The Intergovernmental Panel on Climate Change, working under the auspices of WMO and UNEP and the follow-up activities to the report of the World Commission on Environment and Development are the most prominent and most advanced intergovernmental co-operative endeavours in this regard.

126. The comprehensive and global nature of these approaches suggest that the Principal Subsidiary Bodies of ECE concern themselves with sustainable developments, thereby focusing on specific and regional (east-west) issues falling under their terms of reference. At the same time, they should secure an effective flow of information with each other and with the global programmes mentioned. The efforts of the various Principal Subsidiary Bodies require a co-ordination at the Commission level with a view to establishing a common statistical, analytical and methodological basis (see para. 17 above).

127. Within this framework, the following issues would seem to warrant special attention:

1. internationally-consistent and sustainable long-term energy measures (by means of exchanges of data and views on long-term developments, policy priorities, constraints, remedial measures, implications) (para. 21 above), while taking into account the different development patterns of member States;

2. specific issues of particular relevance to long-term energy and environment sustainability, by

    (a) reducing the projected efficiency "gaps" between ECE countries (paras. 114-118),

    (b) exploring the impact of structural changes of the economy on the energy intensity of the economy,

    (c) accelerating the projected market penetration of renewable sources, especially solar energy, particularly in eastern Europe (paras. 43, 108 (s)),

    (d) diversifying east-west energy relations (paras. 47 and 87),

    (e) in accordance with national laws, regulations and practices, facilitating the transfer of energy efficient and environmentally-acceptable technologies (para. 96),

(f) encouraging international co-operation in the development of energy technologies for the twenty-first century, particularly CO_2 benign and CO_2 control techniques (para. 82).

Notes

1/ World Commission on Environment and Development: Our Common Future, Oxford/New York, 1987.

2/ John G. Hollins/Jane Lagget: Integration of Energy and Environmental Policies in Canada and other OECD Countries. Paper presented to the 14th Congress of the World Energy Conference, Montreal, 1989.

3/ It is recalled that a description of environmental trends and constraints has been elaborated by the Senior Advisers to ECE Governments on Environmental and Water Problems in "Regional Strategy for the Protection of the Environment and Rational Use of Natural Resources in ECE Member Countries covering the period up to the Year 2000 and Beyond", New York, 1988.

4/ J.-R. Frisch, World Energy Horizons: 2000-2020; Technip, Paris 1989.

5/ ECE, Overall Economic Perspective to the Year 2000, New York 1988.

6/ ECE, Energy Balances for Europe and North America, 1970-2000, New York, 1989.

7/ International Energy Agency: Emission controls, Paris, 1988, p. 130.

8/ José Goldemberg, et al., Energy for a Sustainable World, New York 1988.

9/ IEA, op. cit., pp. 103 and 114.

10/ ECE, Convention on Long-Range Transboundary Air Pollution (1979); related Protocol on the Reduction of Sulphur Emissions or their Transboundary Fluxes (1985); related Protocol on NO_x Emissions (1988).

STATISTICAL ANNEX

Table 1

General framework assumptions

	ECE countries (N)	of which:	
		Market economy countries (N1)	Economies of Eastern Europe and USSR (N2)
1. Average annual rate of growth of population			
1973–1985	0.85	0.85	0.80
1985–2000	0.62	0.63	0.60
2000–2010	0.42	0.62	0.49
2. Average annual rate of real growth of GDP, NMP a/			
1985–2000	2.2 – 2.9	2.1 – 2.7	2.4 – 3.5
2000–2010	1.5 – 2.3	1.4 – 2.1	2.0 – 2.9
3. Oil price: dollars per barrel in 1987 prices b/			
1987	18	18	18
2000: moderate growth	25 – 30	25 – 30	25 – 30
low growth	20 – 25	20 – 25	20 – 25
2010: moderate growth	32.50 – 40	32.50 – 40	32.50 – 40
low growth	25 – 30	25 – 30	25 – 30

Source: World Energy Conference, World Energy Horizons 2000–2020, London 1989. GDP (NMP) data were provided to the World Energy Conference by the World Bank and are stated on 1980 United States dollars and exchange rates.

a/ NMP = Net material product: centrally planned economies.

b/ Oil Price – the oil price paths shown for the low and moderate growth scenarios assume the development of GDP/NMP in the ranges shown in the corresponding scenarios.

Table 2

Primary energy demand growth

		1985	2000 Moderate	2000 Low	2010 Moderate	2010 Low
A. PRIMARY ENERGY DEMAND * <u>million toe</u>						
ECE countries	(N)	5 117	6 335	5 917	6 810	6 122
Market economies	(N1)	3 324	4 000	3 701	4 182	3 727
East European economies	(N2)	1 793	2 335	2 216	2 628	2 395
B. AVERAGE ANNUAL GROWTH RATE: (%)						
ECE countries	(N)	.	1.4	1.0	0.7	0.3
Market economies	(N1)	.	1.2	0.7	0.4	0.1
East European economies	(N2)	.	1.8	1.4	1.2	0.8
C. PER CAPITA PRIMARY ENERGY CONSUMPTION* : toe/capita						
ECE countries	(N)	4.72	5.32	4.97	5.45	4.90
Market economies	(N1)	4.82	5.33	4.93	5.34	4.76
East European economies	(N2)	4.55	5.32	5.05	5.63	5.13
D. SAVINGS, a/ in comparison with 1985 due to assumed enhanced energy economy and efficiency: million toe						
ECE countries	(N)	.	1 627	1 182	3 290	2 215
Market economies	(N1)	.	957	839	1 920	1 490
East European economies	(N2)	.	670	343	1 370	725
E. ENERGY INTENSITY* : toe/1000 $ GDP (NMP) (1980 $)						
Market economies	(N1)	0.475	0.384	0.387	0.326	0.340
East European economies	(N2)	0.920	0.715	0.796	0.604	0.706

 * <u>Source</u>: World Energy Conference (WEC), World Energy Horizons 2000-2020, London 1989: the projections are adapted to 2010 as follows: for each region in the WEC study (market and centrally planned economies), an index (with 1985 set at unity) to the projection year 2000 and 2020 is calculated. For 2010, an estimation is made assuming for each energy source that the growth rate estimated for 2000-2020 holds for 2020-2010 as well. This growth index is then applied to the 1985 base year data for market and centrally planned economies using the ECE Energy Data Base. The market economy grouping of the WEC study differs from that of the ECE Energy Data Base in two respects: (a) the WEC study market economy grouping includes Australia, Israel, Japan, New Zealand, and South Africa and additionally (b) non-commercial sources of energy. The presentation in the table above for the market economies is limited to ECE market economies. For the east European economies, the same countries are in the groupings of the WEC study and in the presentation here but the data (as for the market economies) reflect the most recent (small) revisions in the data base for 1985. Non-commercial energy is excluded in both market and east European economies in the ECE Energy Data Base.

 a/   Savings are calculated as the difference between primary energy demand, based on GDP/NMP for 2000 and 2010 multiplied by the energy intensity of 1985 and projected primary energy demand for 2000 and 2010 (from line A above).

Table 3

Primary energy supply growth and pattern * (million toe)

		1985	2000 Moderate	2000 Low	2010 Moderate	2010 Low
SOLID MINERAL FUELS:						
ECE countries	(N)	1 453	1 701	1 581	1 940	1 712
Market economies	(N1)	835	1 006	903	1 131	985
East European economies	(N2)	618	695	677	809	727
PETROLEUM PRODUCTS:						
ECE countries	(N)	1 876	2 085	1 997	2 051	1 925
Market economies	(N1)	1 347	1 505	1 429	1 450	1 341
East European economies	(N2)	529	580	568	601	589
NATURAL GAS:						
ECE countries	(N)	1 217	1 611	1 504	1 661	1 518
Market economies	(N1)	673	765	703	716	646
East European economies	(N2)	544	846	801	946	872
HYDRO POWER: a/						
ECE countries	(N)	287	364	341	402	370
Market economies	(N1)	234	295	277	326	301
East European economies	(N2)	54	69	64	76	69
NUCLEAR POWER:						
ECE countries	(N)	283	543	481	687	567
Market economies	(N1)	236	408	381	510	433
East European economies	(N2)	47	135	100	177	134
NEW ENERGY SOURCES:						
ECE countries	(N)	1	31	13	69	30
Market economies	(N1)	-	21	8	50	21
East European economies	(N2)	1	10	5	19	9
TOTAL (excl. non-commercial energy)						
ECE countries	(N)	5 117	6 335	5 917	6 810	6 122
Market economies	(N1)	3 324	4 000	3 701	4 182	3 727
East European economies	(N2)	1 793	2 335	2 216	2 628	2 395
NON-COMMERCIAL ENERGY:						
ECE countries	(N)	122	120	135	118	142
Market economies	(N1)	79	80	85	80	87
East European economies	(N2)	43	40	50	37	55
TOTAL (incl. non-commercial energy)						
ECE countries	(N)	5 240	6 455	6 052	6 928	6 264
Market economies	(N1)	3 403	4 081	3 786	4 263	3 814
East European economies	(N2)	1 837	2 375	2 266	2 665	2 450

*/ Production plus net imports. See the note to Table 2 for the derivation of total energy supply and for non-commercial energy. For NUCLEAR and NEW ENERGY SOURCES, the data shown are assumed to be the same as for production (Table 6). For the subtotal for SOLID MINERAL FUELS, PETROLEUM PRODUCTS and NATURAL GAS, the shift in share within this grouping is assumed to be the same from 1985 to the projection year as for the corresponding subtotal of the WEC study, calculated separately for the market and centrally-planned economies.

a/ Including steam and NET IMPORTS of electricity.

Table 4

Primary energy supply pattern * (% of total)

		1985	2000		2010	
			Moderate	Low	Moderate	Low
SOLID MINERAL FUELS:						
ECE countries	(N)	28.4	26.8	26.7	28.5	28.0
Market economies	(N1)	25.1	25.2	24.4	27.0	26.4
East European economies	(N2)	34.5	29.7	30.6	30.8	30.2
PETROLEUM PRODUCTS:						
ECE countries	(N)	36.7	32.9	33.8	30.1	31.5
Market economies	(N1)	40.5	37.6	38.6	34.7	36.0
East European economies	(N2)	29.5	24.8	25.6	22.9	24.4
NATURAL GAS:						
ECE countries	(N)	23.8	25.4	25.4	24.4	24.8
Market economies	(N1)	20.2	19.1	19.0	17.1	17.3
East European economies	(N2)	30.4	36.2	36.2	36.0	36.4
HYDRO POWER: a/						
ECE countries	(N)	5.6	5.7	5.8	5.9	6.0
Market economies	(N1)	7.0	7.4	7.5	7.8	8.1
East European economies	(N2)	3.0	3.0	2.9	2.9	2.9
NUCLEAR POWER:						
ECE countries	(N)	5.5	8.6	8.1	10.1	9.3
Market economies	(N1)	7.1	10.2	10.3	12.2	11.6
East European economies	(N2)	2.6	5.8	4.5	6.7	5.6
NEW ENERGY SOURCES:						
ECE countries	(N)	0.0	0.5	0.2	1.0	0.5
Market economies	(N1)	0.0	0.5	0.2	1.2	0.6
East European economies	(N2)	0.1	0.4	0.2	0.7	0.4
TOTAL (excl. non-commercial energy)						
ECE countries	(N)	100.0	100.0	100.0	100.0	100.0
Market economies	(N1)	100.0	100.0	100.0	100.0	100.0
East European economies	(N2)	100.0	100.0	100.0	100.0	100.0

*/   Derived from Table 3.

a/   Including steam and NET IMPORTS of electricity.

Table 5

Net trade by energy source * (million toe)

		1985	2000		2010	
			Moderate	Low	Moderate	Low
SOLID MINERAL FUELS:						
ECE countries	(N)	-12	47	32	56	30
Market economies	(N1)	-31	15	8	20	6
East European economies	(N2)	19	33	24	36	24
PETROLEUM PRODUCTS:						
ECE countries	(N)	-477	-661	-663	-692	-647
Market economies	(N1)	-560	-724	-697	-714	-657
East European economies	(N2)	84	62	34	22	10
NATURAL GAS:						
ECE countries	(N)	-13	-20	-27	-13	-25
Market economies	(N1)	-50	-91	-82	-85	-78
East European economies	(N2)	37	71	55	72	53
HYDRO POWER: a/						
ECE countries	(N)	-	-2	-2	-4	-4
Market economies	(N1)	-1	-4	-2	-6	-3
East European economies	(N2)	1	2	-1	2	-1
NUCLEAR POWER:						
ECE countries	(N)	-	-	-	-	-
Market economies	(N1)	-	-	-	-	-
East European economies	(N2)	-	-	-	-	-
NEW ENERGY SOURCES:						
ECE countries	(N)	-	-	-	-	-
Market economies	(N1)	-	-	-	-	-
East European economies	(N2)	-	-	-	-	-
TOTAL (excl. non-commercial energy)						
ECE countries	(N)	-502	-636	-660	-653	-646
Market economies	(N1)	-643	-804	-772	-785	-731
East European economies	(N2)	141	168	112	132	86

*/ Primary energy production less gross consumption.

a/ Including NET IMPORTS of electricity.

Table 6

Production path */ (million toe)

	1985	2000 Moderate	2000 Low	2010 Moderate	2010 Low
SOLID MINERAL FUELS:					
ECE countries (N)	1 441	1 749	1 612	1 996	1 742
Market economies (N1)	803	1 021	911	1 150	991
East European economies (N2)	638	727	701	846	751
PETROLEUM PRODUCTS:					
ECE countries (N)	1 399	1 423	1 334	1 359	1 279
Market economies (N1)	787	782	732	737	685
East European economies (N2)	612	642	602	623	594
NATURAL GAS:					
ECE countries (N)	1 204	1 592	1 478	1 649	1 493
Market economies (N1)	623	675	621	631	568
East European economies (N2)	582	917	857	1 018	925
HYDRO POWER:					
ECE countries (N)	287	362	339	398	367
Market economies (N1)	233	291	276	320	298
East European economies (N2)	54	71	64	79	69
NUCLEAR POWER:					
ECE countries (N)	283	543	481	687	567
Market economies (N1)	236	408	381	510	433
East European economies (N2)	47	135	100	177	135
NEW ENERGY SOURCES:					
ECE countries (N)	1	31	13	69	30
Market economies (N1)	–	21	8	50	21
East European economies (N2)	1	10	5	19	9
TOTAL (excl. non-commercial energy)					
ECE countries (N)	4 616	5 699	5 257	6 158	5 476
Market economies (N1)	2 682	3 197	2 929	3 398	2 996
East European economies (N2)	1 934	2 502	2 328	2 760	2 481
NON-COMMERCIAL ENERGY:					
ECE countries (N)	123	120	135	118	142
Market economies (N1)	79	80	85	80	87
East European economies (N2)	43	40	50	37	55
TOTAL (incl. non-commercial energy)					
ECE countries (N)	4 738	5 819	5 392	6 275	5 618
Market economies (N1)	2 761	3 277	3 014	3 478	3 083
East European economies (N2)	1 977	2 542	2 378	2 897	2 535

*/   From the market and east European economy groupings of the WEC study, an index (with 1985 set at unity) is calculated and applied to the corresponding market and east economy groups from the ECE Energy Data Base. For production by energy source, the shift in shares from 1985 to the projection year in the country groupings of the WEC study is assumed to apply equally to the country groupings using the ECE Energy Data Base. The projections for non-commercial sources of energy are taken from the WEC study as follows: for the market economies, 9.5 million toe of non-commercial energy attributed to non-ECE market economies in the 1985 base year is subtracted from the market economy countries of the WEC study to estimate the production of non-commercial energy for the ECE market economies in 1985. The same ratio of non-commercial energy for ECE vs. NON-ECE market economies is also assumed to apply to other years.

Table 7

Production pattern */ (% of total)

	1985	2000		2010	
		Moderate	Low	Moderate	Low
SOLID MINERAL FUELS:					
ECE countries (N)	31.2	30.7	30.7	32.4	31.8
Market economies (N1)	30.0	31.9	31.1	33.9	33.1
East European economies (N2)	33.0	29.1	30.1	30.6	30.3
PETROLEUM PRODUCTS:					
ECE countries (N)	30.3	25.0	25.4	22.1	23.3
Market economies (N1)	29.3	24.5	25.0	21.7	22.9
East European economies (N2)	31.7	25.6	25.9	22.6	23.9
NATURAL GAS:					
ECE countries (N)	26.1	27.4	28.1	26.8	27.3
Market economies (N1)	23.2	21.1	21.2	18.6	19.0
East European economies (N2)	30.1	36.7	36.8	36.9	37.3
HYDRO POWER:					
ECE countries (N)	6.2	6.3	6.5	6.5	6.7
Market economies (N1)	8.7	9.1	9.4	9.4	9.9
East European economies (N2)	2.8	2.8	2.7	2.9	2.8
NUCLEAR POWER:					
ECE countries (N)	6.1	9.5	9.1	11.2	10.4
Market economies (N1)	8.8	12.7	13.0	15.0	14.4
East European economies (N2)	2.4	5.4	4.3	6.4	5.4
NEW ENERGY SOURCES:					
ECE countries (N)	0.0	0.5	0.2	1.1	0.5
Market economies (N1)	0.0	0.7	0.3	1.5	0.7
East European economies (N2)	0.1	0.4	0.2	0.7	0.3
TOTAL (excl. non-commercial energy)					
ECE countries (N)	100.0	100.0	100.0	100.0	100.0
Market economies (N1)	100.0	100.0	100.0	100.0	100.0
East European economies (N2)	100.0	100.0	100.0	100.0	100.0

*/   Derived from Table 6.

Table 8

Conventional primary energy reserves and use

	COAL	OIL	GAS	HYDROPOWER	URANIUM
	10*9 toe	10*9 toe	10*9 toe	TWh million toe	10*3 tons million toe
A. WORLD					
(a) Proven recoverable conventional reserves	896	97	74	19 000 TWh	3 321 10*3 t
(b) Estimated additional conventional resources	2 699	37	157	...	3 999 10*3 t
(c) Assumed cumulative production 1985-2010 of conventional (and in the case of oil: unconventional) sources under					
(i) moderate economic growth	91	104	58	20 mill. toe	28 mill. toe
(ii) low economic growth	78	86	49	18 mill. toe	21 mill. toe
B. ECE COUNTRIES					
(a) Proven recoverable conventional reserves	354	17	42	...	1 371 10*3 t
(b) Estimated additional conventional resources	1 993	3	130	...	2 738 10*3 t
(c) assumed cumulative production 1985-2010 of conventional and in the case of oil: unconventional) sources under					
(i) moderate economic growth	44	36	38	9 mill. toe	13 mill. toe
(ii) low economic growth	41	35	36	9 mill. toe	11 mill. toe

Sources and methods:

For RESERVES - COAL, OIL, GAS, URANIUM:

Jean-Romain Frisch, Future Stresses for Energy Resources, World Energy Conference, 1986.

The RESERVES data for coal, oil, gas and uranium for the ECE countries also include ISRAEL.

For RESERVES - HYDRO:

H. Blind, Weltweite Nutzung der Wasserkräfte, Energiewirtschaftliche Tagesfragen 6/1988.

The RESERVES data for the world exclude USSR, Eastern Europe and China.

For CUMULATIVE PRODUCTION:

Jean-Romain Frisch, op. cit., Annex 7, projections H and C respectively (for moderate and low growth scenarios). The data are reported for 1985-2000 and 2000-2020. Interpolation is used to approximate cumulative production for 2000-2010. For the ECE countries, the projections are made in connection with the annual production projections for 1985, 2000 and 2010 assuming a constant compound annual growth rate between the two pairs of benchmark years 1985-2000 and 2000-2010.

Note: The COMPARABILITY OF ESTIMATES of energy reserves and resources between fuels, areas and periods suffers from differences in the geological, economic and technological assumptions used. The data therefore provide only orders of magnitude.

Table 9

New and renewable sources of energy (million toe)

	1985	2000		2010	
		Moderate	Low	Moderate	Low
ECE countries (N)					
Total	411	513	487	585	539
- hydropower	287	362	339	398	367
- new sources	1	31	13	69	30
- non-commercial energy	123	120	135	118	142
Market economies (N1)					
Total	312	392	369	450	406
- hydropower	233	291	276	320	298
- new sources	–	21	8	50	21
- non-commercial energy	79	80	85	80	87
East European economies (N2)					
Total	98	121	119	135	133
- hydropower	54	71	64	79	69
- new sources	1	10	5	19	9
- non-commercial energy	43	40	50	37	55

New and renewable sources of energy (% of total primary energy supplies)

	1985	2000		2010	
		Moderate	Low	Moderate	Low
ECE countries (N)					
Total	8.7	8.8	9.0	9.3	9.6
- hydropower	6.1	6.2	6.3	6.3	6.5
- new sources	–	0.5	0.2	1.1	0.5
- non-commercial energy	2.6	2.1	2.5	1.9	2.5
Market economies (N1)					
Total	11.3	12.0	12.2	12.9	13.2
- hydropower	8.4	8.9	9.2	9.2	9.7
- new sources	–	0.6	0.3	1.4	0.7
- non-commercial energy	2.9	2.4	2.8	2.3	3.5
East European economies (N2)					
Total	5.0	4.8	5.0	4.7	5.2
- hydropower	2.7	2.8	2.7	2.7	2.7
- new sources	–	0.4	0.2	0.7	0.4
- non-commecial energy	2.2	1.6	2.1	1.3	2.2

PART II

RESEARCH NOTES

CO_2-CONCENTRATION AND ENERGY SCENARIOS

Purpose

1.   Chapter III above describes the impact of rising CO_2 concentrations on the environment in qualitative terms. The present monograph aims at quantifying the CO_2 implications of recent world energy demand/supply projections of which the ECE projections described in chapter II are the most important component. A key concern is that these anticipated energy developments and their continuation beyond the year 2010 will likely result in a doubling of the pre-industrial level of CO_2 concentrations (taken to be from 275 parts per million (ppm) to 550 ppm) before the end of the next century. ' Taking a normative approach to the problem, this monograph also describes a global energy strategy which delays the doubling of pre-industrial CO_2 concentrations into the twenty-second century and suggests supporting measures.

Methodological base

2.   Extending the ECE energy demand/supply scenarios shown in tables 3 and 6 of the Statistical Annex, alternative scenarios were specified for energy developments, carbon emissions of fossil fuels and development of the CO_2 concentration levels employing the projections framework presented by William D. Nordhaus and Gary W. Yohe in _Climate Change, Report of the Carbon Dioxide Assessment Committee_, National Academy Press, 1983 and applied by Yohe in a consultant's report to the fourteenth Congress of the World Energy Conference, "Notes on the Environmental Consequences of the 'New Energy Perspectives'", March 1989. 1/

3.   The projections equation is a dynamic representation of the CO_2 accumulation process. Changes in accumulated carbon dioxide concentration (stated in parts per million of carbon) in a given time period (t) are related linearly to emissions during a given time interval (E(t)) and the prior period's CO_2 concentration level (AC(t-1)). Multiplying the first variable E(t) are two coefficients, a conversion factor to translate emissions in gigatons of carbon into parts per million of CO_2 in the atmosphere (0.471) and a second coefficient, the marginal airborne fraction of carbon remaining in the atmosphere (per year) per unit of emissions. This latter coefficient is not yet known with precision and it is said that the marginal airborne coefficient might range from 0.38 to 0.59. 2/ The coefficient applied in the scenario exercises is 0.47 following Nordhaus and Yohe. The coefficient multiplying the prior period's level of atmospheric concentration of CO_2 is a negative seepage factor (-.001) which proximately represents the capacity of the earth to absorb carbon from the atmosphere per time period.

4.   It will be noted that the calculations exclude greenhouse gases other than CO_2. Should these, particularly the CFC's remain significant, climate variations would occur earlier than anticipated on the basis ONLY of rising CO_2 concentrations. Also, the calculations refer to global rather than ECE indications not only for the reasons cited above but additionally because of the difficulty in calculating regional climate variation that may be brought about by rising concentrations of CO_2 and other greenhouse gases.

Questions addressed

5.   For the purposes of analysis, a number of scenarios were constructed.
Each of these scenarios features an assumed path for the development of
world fossil fuel demand and the resulting rate of carbon emissions and
CO_2 concentrations. For the purposes of comparison, each of the scenarios
may be referred to a "Baseline Scenario" which is taken to be the fossil
fuel demand path specified for the "moderate economic growth scenario"
to 2020 in World Energy Conference, Global Energy Perspectives 2000-2020,
fourteenth Congress of the World Energy Conference (Montreal, 1989) and
extended by Yohe to 2080 in his consultant's report cited above. The paths
for fossil fuel use, carbon emissions (and their growth rates for the period
shown) and CO_2 concentrations specified are as follows:

Baseline scenario

Year	Fossil fuel use	Emissions		CO_2 concentrations
	10*3 mtoe	GT carbon	Growth rate	ppm
1985	6.0	5.50	–	333
2000	7.9	7.23	1.84	349
2020	9.9	9.12	1.17	377
2040	12.5	10.58	0.79	410
2060	16.5	13.97	1.44	456
2080	18.6	18.23	1.34	511

The baseline scenario exhibits CO_2 concentrations that steadily rise and
surpass 500 ppm by the last quarter of the next century. The annual increment
to the CO_2 level is about 3.5 ppm in 2080. Even if the emissions were to be
stabilized at their 2080 rate, a doubling of CO_2 concentrations over the
pre-industrial level would occur before the end of the twenty-first century.

6.   The framework of analysis may be used to specify other scenarios.
Here, it was decided to choose a number of alternatives in which different
paths for emission rates could be considered and the consequences drawn for
CO_2 concentrations. Accordingly three projections were carried out each
year from 1985 to a selected date - 2030, 2060 and 2090. The object of
these (first) three scenarios of the exercise was to ascertain for the
years mentioned the compound annual growth rate of emissions at which
CO_2 concentrations would reach double the pre-industrial level of
concentration. In the fourth scenario it is assumed that emissions grow
at a 1% compound annual rate from 1985 to 2000 but then decline at rates of
1 to 3% per annum respectively in three variants of this scenario. The object
of this fourth scenario is to find the rate of decline from 2000 onwards
(and the associated level of emissions) that would permit a (relative)
leveling off in the CO_2 concentration level. Tables 1 to 4 correspond
to the four scenarios. Finally, a "Preferred (or normative) Scenario"
(shown in table 5) is considered which, while it does not assume a
stabilization of CO_2 concentrations in the twenty-first century, "buys
sufficient time" for future generations to find new ways of achieving an, as
yet, undetermined technical breakthrough in CO_2 removal. The Nordhaus and
Yohe projections equation was applied to a 1985 world level of CO_2
concentrations of 333 ppm and an emission rate of 5.5 gigatons of carbon
emissions from fossil fuels given in Yohe, op. cit. 3/

Scenario 1. Doubling of CO_2 concentrations by 2030

7.   The first scenario shows emission growth rates and associated concentration levels for alternative compound annual growth rates of emissions of 3.0, 3.5, 4.0, 5.0, 5.4 and 5.5%. For simplicity of presentation, it is assumed in the first four scenarios that the relative shares of coal, oil and gas within the total for fossil fuel demand are constant throughout the projections period (an assumption that will be altered to reflect projections for these fuels given in the "Preferred Scenario"). 4/ It is seen that the 550 ppm barrier is reached by 2030 only when the growth rate of emissions is set between 5.4 and 5.5% per annum. If the Nordhaus and Yohe values of the basic parameters are close to the "true" values and also assuming that the framework of analysis used is a reasonable representation of what admittedly is a highly complex process, IT WOULD SEEM HIGHLY UNLIKELY THAT A DOUBLING OF THE PRE-INDUSTRIAL ATMOSPHERIC CONCENTRATION OF CO_2 WOULD TAKE PLACE AS SOON AS 2030. 5/ This rate would require both extremely rapid (and presently unanticipated) increases in overall energy use and/or a shift towards carbon-based energy, particularly solid fuels. Thus scenario 1 with a doubling of the pre-industrial level of CO_2 concentrations in about 30 years would seem to be an unlikely event.

Scenarios 2 and 3. Doubling of CO_2 concentrations by 2060 and 2090

8.   In the second scenario, a 2.3 to 2.4% compound annual growth rate of fossil fuel demand is assumed. With a much lower growth rate of fossil fuel demand, the rate of carbon emissions rise much more slowly than in the first scenario. The resulting impact on CO_2 concentrations is more gradual and the 550 ppm barrier is not reached until 2060 (see table 2). In the third scenario, an even lower growth rate is chosen, between 1.2% and 1.3% and, as would be expected, and the rise in carbon emissions is correspondingly even more gradual. Here, carbon emissions do not reach the 19-21 gigaton/yr rate until 2090, about a half century later than in scenario 2. It takes about 30 years more in this instance, compared with scenario 2, for the doubling of pre-industrial CO_2 levels to take place.

9.   THE LESSON OF (THE BASELINE AND) THE FIRST THREE SCENARIOS IS CLEAR - EVEN LOW BUT STEADY GROWTH RATES OF EMISSIONS INEVITABLY DRIVE UP, YEAR BY YEAR, THE CONCENTRATION LEVELS OF CO_2 SO THAT EVENTUALLY - NO DOUBT BEFORE THE TURN OF THE TWENTY-SECOND CENTURY - CO_2 CONCENTRATION REACH LEVELS THAT ARE NOW THOUGHT TO RISK IMPORTANT (AND DANGEROUS) CLIMATE CHANGES. THE PROBLEM MAY NOT BE FACED BY THE CURRENT GENERATION BUT IS HIGHLY LIKELY TO BE CONFRONTED BY FOLLOWING GENERATIONS.

Scenario 4. Relative stabilization of CO_2 concentrations by 2090

10. The fourth scenario is built on the premise that if carbon emissions rates are cut substantially enough, CO_2 concentrations can be stabilized. There is no one path that describes the scenario conditions. One might choose to allow emission rates to rise and then fall, rise and stabilize or some other combination of events. The growth rates of these events can be set in numerous combinations. Shown in table 4 are three such combinations. Each of the variants assumes that emissions rise at a 1% compound annual rate from 1985 to 2000 but decline thereafter, at alternative rates of 1 to 3%. It is seen that only when the rate of decline is between 2 and 3% per annum would the concentration level of CO_2 (roughly) stabilize by 2090. The

required rate of emissions associated with this event is under half of the 1985 rate. Clearly, such an outcome would require both a significant success in reducing energy use per unit of GDP (NMP) at rates faster than GDP (NMP) growth rates and a reduction in the share of world usage of carbon-based fuels (particularly coal) in overall energy use. This would require an (as yet) unforeseen rise in the share of other energy sources as a compliment to reduction in energy intensities. 6/

11. In addition to substitution of other energy sources against carbon-based fuels and away from coal to less-carbon intensive fuels within the carbon-based fuels, there is the possibility that future technology may well play a positive role in reducing carbon emissions in stationary power sources. A number of recent developments are reported in I. Smith in CO$_2$ and Climatic Change, IEA Coal Research (1988). This report reviews ongoing efforts to improve the combustion cycle (burning coal in CO_2/O_2 rather than air), improved removal of CO_2 from flue gases and disposal of CO_2. Smith estimates that CO_2 control in United States power plants alone has the potential of reducing "CO_2 concentration in the atmosphere by 10%". 7/ These possibilities referred to above suggest that there is some room for a reduction in the effective emission rate per unit of fossil fuel consumed. Smith also notes the important contribution that a world-wide reforestation programme might make to CO_2 removal.

A preferred (normative) scenario

12. An additional, normative or preferred scenario has been designed which, while it does not assume a stabilization of CO_2 concentrations in the twenty-first century, "buys sufficient time" for future generations to find new ways of achieving, as yet, undetermined technical progress in CO_2 removal. The main features of the scenario are shown in table 5. The formal conditions imposed in this preferred scenario are:

(a) That CO_2 concentrations above 400 ppm do not occur until about 2030;

(b) That fossil fuel demand and hence, carbon emissions, stabilize as of 2020; and

(c) That the energy strategy response is evolutionary, balanced and global, rather than disruptive, selective or regional.

This latter response relies heavily on a further increase of energy efficiency from 0.49 mtoe per billion (1980) dollars of world GDP in 2000 to half that level in 2040. This order of reduction is predicated as non-fossil fuel sources of energy must supply all incremental requirements for energy beyond 2020. A second assumption, serving to reduce carbon emissions within fossil fuel use, stipulates an increased share of gas and a decreased share of solid fuels. Specifically, the share of gas is augmented by 3% (10%) in 2000 (2020) of the world total for energy demand and that of solid fuels reduced by corresponding amounts. The pattern of fossil fuel use for 1985-2020 in this preferred scenario compared with the baseline scenario is as follows:

World primary energy use of fossil fuels (mtoe)

Year	Baseline scenario				Preferred scenario			
	Fossil fuels	Coal	Oil	Gas	Fossil fuels	Coal	Oil	Gas
1985	6 001	2 116	2 497	1 388	6 001	2 116	2 497	1 388
2000	7 885	2 816	3 088	1 981	7 885	2 508	3 088	2 289
2020	9 949	4 051	3 543	2 355	9 949	2 699	3 543	3 707
%								
1985	100.0	35.3	41.6	23.1	100.0	35.3	41.6	23.1
1985	100.0	35.7	39.2	25.1	100.0	31.8	39.2	29.0
1985	100.0	40.7	35.6	23.7	100.0	27.1	35.6	37.3

The preferred scenario results in the following fossil fuel emission and CO_2 concentration path over the scenario period to 2090.

Preferred scenario: fossil fuel use, emissions and CO_2 concentrations

Year	Fossil fuel use	Emissions	CO_2 concentrations
	10*3 mtoe	GT carbon	ppm
1985	6.0	5.50	333
2000	7.9	7.04	349
2020	9.9	8.57	376
2040	9.9	8.57	406
2060	9.9	8.57	436
2080	9.9	8.57	465
2090	9.9	8.57	479

13. The results of the above assumptions are for fossil fuel carbon emissions to reach 8.57 gigatons in 2020 and remain at that rate thereafter. CO_2 concentrations exceed 400 ppm only in the middle of the twenty-first century, rising from 2040 to 2090 at an average annual rate of 1.46 ppm per year. 8/

14. The preferred scenario being a normative one reflects the judgement of its authors as to the urgency, necessary emphasis and plausibility of a global response to the risk of global warming. There can be numerous other normative scenarios, according to policy postures. The preferred scenario is thought to advance the discussion on global energy response strategies in the following respects:

(a) It shows quantitatively that an evolutionary adaptation of the present world energy system can buy sufficient time for lasting solutions to be found to the reduction, removal and disposal of CO_2, while admitting an increase of CO_2 concentrations of about 1.75 the pre-industrial level of 275 ppm.

(b) In using all options (acceptance of a significant increase of CO_2 concentration; abatement strategies both on the energy demand and supply side), the preferred scenario appears to be "least painful" among the adaptation scenarios presented here. In particular, it does not assume excessive reliance on solar or nuclear power or a reduction in the use of coal (here solid fuel consumption is 28% higher in 2040 than in 1985). Indeed the normative scenario features rises in all energy sources but with preference given to non-fossil sources of energy and within fossil fuel sources, the least carbon emitting.

(c) The preferred scenario is considered compatible with the quest for a growing economic welfare of a rising population, with energy reserves and resources, with available or emerging best technologies and with financing possibilities.

15. However, as for every world model, the preferred scenario may be applied only with difficulty at the regional level. An example of such a difficulty is coal with growth limited to 583 mtoe in the preferred scenario rather than 1,935 mtoe given in the moderate growth reference scenario of the World Energy Conference. The burden of the slow-down in coal production and consumption may well have to be borne disproportionately by the ECE region and other industrial countries if the centrally planned Asian countries (China especially) insist on increasing coal production and use by the 774 mtoe anticipated in the reference scenario. All this underlines the implicit assumption of the preferred scenario that the solution to the issue of global warming as to world-wide co-operation and mutual accommodation.

Tentative conclusions

16. The energy demand/supply scenarios for the world (and the ECE) discussed in the present study would sooner or later, despite an assumed significant rise of energy efficiency, drive up world concentration of CO_2 to levels that are thought to risk a significant rise in global temperatures with attendant risks. The attainment of lower growth rates of carbon emissions as proposed in the "Preferred Scenario" would be favourable as it would postpone into the twenty-second century the date at which a doubling of CO_2 concentrations above the pre-industrial level would take place. The time gained would permit the development of alternatives to carbon-based fuels, technological improvements in the use of carbon-based fuels, interfuel substitution within carbon-based fuels and finally CO_2 removal measures.

The role of other greenhouse gases

17. While it is frequently stated that the major determinant of future climate change is CO_2 (and that the CO_2 problem is largely an energy use problem), it is now recognized that there is a reinforcing role played by a number of other greenhouse gases, sometimes termed radiatively important gases (RIGS). 9/ The key RIGS (other than CO_2) are said to be methane, nitrous oxide and CFC-11 and CFC-12 (chlorofluorocarbons). Increases in methane

concentration are linked to increases in human populations; nitrous oxide, primarily with fossil fuel combustion and natural soil releases; the chlorofluorocarbons (CFCs), with industrial processes. The overall impact of increases in (non-CO_2) RIGS is said to be to reinforce the impact of rising CO_2 concentrations. In Changing Climate, op. cit., the estimated impact increases in the concentration of RIGS and non-RIGS trace gases was said to be potentially as important as CO_2 (assuming the order of increase of these gases and that these increases took place in all of the gases). 10/

18.   Taking into account the principal gases contributing to the greenhouse effect, a recent study by the United States Environmental Protection Agency, Policy Options for Stabilizing Global Climate, Draft Report to Congress (February 1989), focuses on CO_2, methane, nitrous oxide and chlorofluorocarbons (particularly CFC-11 and CFC-12) as the key gases. The weights of these gases in contributing to global warming are said to be as follows: CO_2 (49%), methane (18%), nitrous oxide (6%), CFC-11 and CFC-12 taken together (14%). This leaves a weight of 13% for other greenhouse gases which include other halons, tropospheric ozone and stratospheric water, though it is stated that the contribution of the "other" category is uncertain.

19.   The EPA Report discusses the interactions among greenhouse and non-greenhouse gases, pointing out that the effect of any one greenhouse gas depends on the concentration of other gases. This makes long-term climatic projection work highly conjectural. Additionally, the EPA Report emphasizes that CO_2 emissions are not only the result of energy consumption alone but HOW energy is consumed - one example being the output of CO_2 in cement manufacture. Also the important role deforestation has in either contributing to or reducing CO_2 emissions and absorption is treated in addition to other elements of land use in their relation to CO_2 and other greenhouse gases. The EPA projections for carbon emissions take into account both primary energy demand for coal, oil, gas and ADDITIONALLY other sources. It is the "other sources" element that accounts for an additional 0.4 gigatons of carbon emissions per annum thus raising the total emission level for 1985 from 5.5 gigatons from fossil fuels only (used by Yohe, op. cit.) to 5.9 gigatons for all sources used in the projections work of the EPA. It is also notable that the EPA projections to the year 2100 cover ALL MAJOR GREENHOUSE GASES and not just CO_2. This procedure allows for a comprehensive evaluation of the possible climatic effects of greenhouse gases (when the weights of the greenhouse gases in combination with their concentrations are put together).

20.   An application of the findings of the EPA Report to the scenarios presented above implies a probable acceleration of the EQUIVALENT CO_2 concentrations of the order of (perhaps) one or two decades. It enhances the dramatic significance of greenhouse gases as the MAIN factor threatening the sustainability of the world's (and ECE's) energy developments.

TABLE 1 (Scenario 1)

Alternative fossil fuel emission growth rates
and accumulated global CO_2 concentrations a/

Year	(a) 3.0% growth of emissions		(b) 3.5% growth of emissions	
	Emission rate	CO_2 concentrations	Emission rate	CO_2 concentrations
1985	5.5	333.0	5.5	333.0
2000	8.6	351.2	9.2	352.2
2010	11.5	370.0	13.0	373.4
2020	15.5	396.3	18.3	404.4
2030	20.8	432.6	25.9	449.5

Year	(c) 4.0% growth of emissions		(d) 3.5% growth of emissions	
	Emission rate	CO_2 concentrations	Emission rate	CO_2 concentrations
1985	5.5	333.0	5.5	333.0
2000	9.9	353.3	11.4	355.5
2010	14.7	377.0	18.6	385.2
2020	21.7	413.6	30.3	435.6
2030	33.1	469.2	44.4	519.6

Year	(c) 5.4% growth of emissions		(d) 5.5% growth of emissions	
	Emission rate	CO_2 concentrations	Emission rate	CO_2 concentrations
1985	5.5	333.0	5.5	333.0
2000	12.1	356.5	12.3	356.7
2010	20.5	388.9	21.0	389.9
2020	34.7	446.0	35.8	448.8
2030	58.6	544.8	61.2	551.6

Result: A rise to 550 ppm CO_2 concentration by year 2030 occurs with a 5.4% to 5.5% growth rate of emissions.

a/ Emission rates are in gigatons of carbon; CO_2 concentrations in ppm.

TABLE 2 (Scenario 2)

Alternative fossil fuel emission growth rates and accumulated global CO_2 concentrations a/

Year	(a) 2.0% growth of emissions		(b) 2.3% growth of emissions	
	Emission rate	CO_2 concentrations	Emission rate	CO_2 concentrations
1985	5.5	333.0	5.5	333.0
2000	7.4	349.4	7.7	349.9
2020	11.0	382.7	12.2	386.5
2040	16.3	434.9	19.2	447.3
2060	24.3	515.2	30.3	546.4

Year	(c) 2.4% growth of emissions	
	Emission rate	CO_2 concentrations
1985	5.5	333.0
2000	7.9	350.1
2020	12.6	387.8
2040	20.3	451.8
2060	32.6	558.0

Result: A rise to 550 ppm CO_2 concentration by year 2060 occurs with a 2.3% to 2.4% growth rate of emissions.

a/   Emission rates are in gigatons of carbon; CO_2 concentrations in ppm.

TABLE 3 (Scenario 3)

Alternative fossil fuel emission growth rates and accumulated global CO_2 concentrations a/

Year	(a) 1.0% growth of emissions		(b) 1.2% growth of emissions	
	Emission rate	CO_2 concentrations	Emission rate	CO_2 concentrations
1985	5.5	333.0	5.5	333.0
2000	6.4	347.7	6.6	348.0
2030	8.6	386.4	9.4	389.9
2060	11.6	441.0	13.5	452.9
2090	15.6	516.9	19.2	546.1

Year	(c) 1.3% growth of emissions	
	Emission rate	CO_2 concentrations
1985	5.5	333.0
2000	6.7	348.2
2030	9.8	391.7
2060	14.5	459.3
2090	21.3	562.4

Result: A rise to 550 ppm CO_2 concentration by year 2090 occurs with a 1.2% to 1.3% growth rate of emissions.

a/ Emission rates are in gigatons of carbon; CO_2 concentrations in ppm.

TABLE 4 (Scenario 4)

Alternative fossil fuel emission growth rates and accumulated global CO_2 concentrations a/

Year	(a) Emissions grow by +1% for 1985-2000 and -1% for 2000-2090	
	Emission rate	CO_2 concentrations
1985	5.5	333.0
2000	6.4	347.7
2090	4.7	446.6

Year	(b) Emissions grow by +1% for 1985-2000 and -2% for 2000-2090	
	Emission rate	CO_2 concentrations
1985	5.5	333.0
2000	6.5	347.7
2060	4.7	416.7
2090	2.6	427.2

Year	(c) Emissions grow by +1% for 1985-2000 and -3% for 2000-2090	
	Emission rate	CO_2 concentrations
1985	5.5	333.0
2000	6.4	347.7
2060	3.5	411.3
2090	1.9	414.3

Result: A rate of decline of 2% - 3% after 2000 is required to (relatively) stabilize CO_2 concentrations (with small annual increments at the end of the period considered).

a/ Emission rates are in gigatons of carbon; CO_2 concentrations in ppm.

TABLE 5

Preferred Scenario

	1985	2000	2020	2040	2060	2080	2090
GDP (billions of 1980 $)	12 944	20 835	36 120	62 628 a/	...	...	...
Total Energy Demand (mtoe)	7 669	10 259	13 525	15 419	...	...	...
Total Energy Demand/GDP	0.5925	0.4924	0.3744	0.2462 b/	...	...	...
Fossil Fuel Demand:	6 001	7 885	9 949	9 949	9 949	9 949	9 949
- Coal	2 116	2 508	2 699	2 699	...	...	...
- Oil	2 497	3 088	3 543	3 543	...	...	...
- Gas	1 388	2 289	3 707	3 707	...	...	...
Non-Fossil Fuel Demand:	1 668	2 374	3 576	5 470	...	...	...
- Hydro	445	642	1 043	1 826	...	...	...
- Nuclear	324	637	1 113	1 950	...	...	...
- New and renewable sources	19	70	365	639	...	...	...
- Non-commercial	880	1 025	1 055	1 055	...	...	...
Fossil Fuel Emissions (GT):	5.499	7.044	8.570	8.570	8.570	8.570	8.570
- Coal	2.444	2.897	3.117	3.117	3.117	3.117	3.117
- Oil	2.163	2.675	3.069	3.069	3.069	3.069	3.069
- Gas	0.892	1.472	2.384	2.384	2.384	2.384	2.384
CO_2 concentration (ppm)	333.0	348.8	376.2	406.3	435.9	464.8	479.0

a/   Based on a 2.79% compound annual growth rate for 2020-2040 world GDP.

b/   Assumed to be one half the year 2000 total energy demand/GDP ratio. GT: gigatons.

FIGURE 1

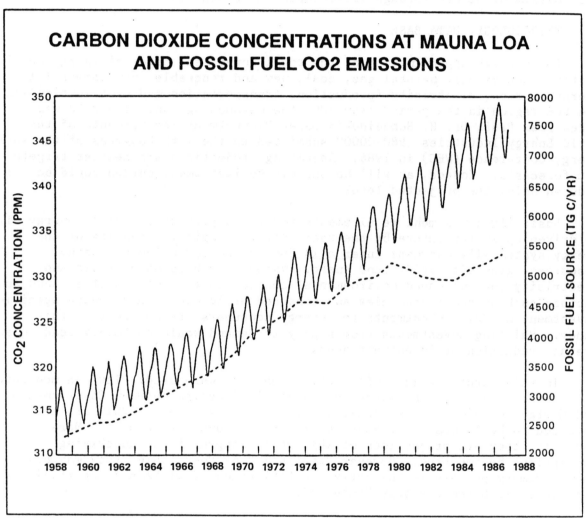

CARBON DIOXIDE CONCENTRATIONS AT MAUNA LOA AND FOSSIL FUEL CO2 EMISSIONS

Source: EPA Report, op. cit., pp. 1-10.

INVESTMENT REQUIREMENTS OF THE ENERGY SUPPLY INDUSTRIES
IN THE ECE REGION, 1980-2000

A. PURPOSE

1.   The projected increase of indigenous energy production in the ECE region
during 1985-2000 (14-24%, table 6) would require substantial investment of the
(capital-intensive) energy supply industries. The purpose of the present note
is to explore the orders of magnitude involved and possibilities for reducing
the investment burden through all-European co-operation.

B. METHODOLOGY, DATA BASE

2.   For a rough approximation of the total investment needs of energy supply
sectors such as oil, natural gas, coal, new and renewable, non-commercial
energy sources, electricity (production, transportation and distribution) in
the ECE region in the period 1980-2000 the methodology and some data were
taken from Prof. Dr. H. Schneider's paper "Investment requirements of the
World Energy Industries 1980-2000" submitted to the 13th Congress of the World
Energy Conference (WEC) in 1986. Resulting projections are neither targets
nor forecasts. Indeed, as will be shown, the investment burden could be
reduced below the projected level.

3.   Basically investment requirements for the supply of a unit of energy will
be driven up in the future by the fact that the capital intensity of the
energy system will increase _inter alia_ due to the depletion of energy
resources, growth of grid-bound energy systems, such as natural gas and
electricity and increased environmental control costs. On the other hand some
factors such as scale economies and enhanced efficiency would reduce specific
investment needs. Investments for energy consumers are not part of this
study. Shifting investments from supply to demand would certainly lead to an
overall reduction of investment needs.

4.   In this paper two economic growth scenarios were used: low and moderate,
which were taken from the study on "Interrelationships between Environmental
and Energy Policies". The scenarios for the years 2000 and 2010 are not
intended to be forecasts of what is likely, or ought, to happen. More
modestly, they are projections of what would happen if the underlying
assumptions became true. However "realistic" or "desirable" they may be,
their primary purpose is the early identification of risks and opportunities
associated with these energy "futures".

5.   Concerning the data base, Dr. Schneider's projections have been used to
estimate specific investments per unit of energy supplied; projections of
production of primary energy sources have been taken from the statistical
annex. Since this statistical annex does not contain estimates of electricity
generation, the present research note contains estimates of the percentage
share of total primary energy consumption used for electricity generation,
which is assumed to increase between 1970 and 2020 from 25% to 44% in the ECE
region and from 25% to 50% in North America and Western Europe, from 24% to
34% in the USSR and Eastern Europe (see table 1).

- total electricity generation (TWh) is derived from total energy input in electricity generation assuming an efficiency of about 34% (tables 2, 3, 4, 5); the consequent estimations of electric generating capacity are calculated using load factors (tables 3, 4, 5);

- gross additions to electric generating capacity during 1980-2000 are the sum of replacement capacities and net additions to generating capacity (table 6); these are calculated on the assumption of constant yearly rates of retirements as well as net additions in 1980-2000. As can be seen, there are significant differences, for the various energy sources, in the estimates of per unit investment requirements for electricity generation, transportation and distribution (tables 7, 8).

- non-commercial energy sources (NCE); in 1980, in the centrally planned economies (CPE) part of non-commercial energy sources (NCE) was used for electricity production. Per unit investment for production of 1 toe is estimated $150-160. Total investment requirements for electricity production, transportation and distribution are about $300/toe. In the market economies (ME) only a small part of cumulative production of non-commercial energy sources has and would be utilized for electricity production. The rest is used for water heating, drying, in gasifiers, etc. Installed plant cost for gasifiers are estimated to be approximately $15/toe. 11/ Cumulative production of non commercial energy over the period 1980-2000 is estimated at 850-952 mtoe (eastern Europe) and 1613-1658 mtoe (MEs).

- new and renewable energy sources in the ECE region such as solar, wind, geothermal, mini-hydro mostly are utilized for electricity production. Part of solar and geothermal energy is applied for heating purposes (solar passive, solar pond, solar flat plate collectors, solar cookers, geothermal direct heat). Per unit investment requirements of new energy sources are given in table 8. Investment requirements for NCE and new energy sources used for electricity production are included in column electricity (table 11). Environmental costs are not included in this paper but if they were, the total investment requirements would be substantially higher. If only all coal-fired plants were equipped with emission control technologies, it would require perhaps more than 300 billion (1980 US) dollars or 12-14% of the total cumulative investment requirements for electricity generation in the ECE region over the period 1980-2000.

C.   ESTIMATED INVESTMENT GROWTH AND PATTERN

(a)   Growth rates

6.   During 1980-2000 the ECE countries are projected to increase energy investment requirements by 22-40% (according to whether a low or moderate economic growth scenario applied). This is an average annual growth rate of 1-2% (see table 10).

7. In the market economies projected expenditure growth over this period would be 16-34%. The share of these countries in the total investments of the energy industries in the ECE region would gradually decline from 76% in 1980 to 68% in 2000 (moderate scenario). The same tendencies apply to oil (62% and 52.6%); gas (94% and 90.6%); coal (81.6% and 78.4%) and electric power (77% and 68%) respectively.

8. By contrast, in eastern Europe, the total growth of investments requirements would be 39-97% over the period with a rising share in ECE energy investments: from 24% (1980) to 32% (2000).

(b) Pattern

9. In 1980 about half of the total investment requirements of the energy supply industries in the ECE region was for energy production (including generation of electricity) and half for energy transportation and distribution. By 2000 investment requirements in energy production are expected to grow somewhat faster than investments in transportation and distribution.

10. A breakdown by energy sources indicates that investment requirements are and would be dominated by the electricity sector (more than 2/3 of the total).

(c) Cumulative investments 1980-2000

11. The cumulative investment requirements of ECE energy supply industries during 1980-2000 will be approximately 7,300 billion (1980 US) dollars (low scenario) and 8,900 billion (1980 US) dollars (moderate scenario). Out of this ECE total, $US 5,500 and 6,500 billion would be accounted for by the market economy countries respectively and $US 1,800 and 2,400 billion by eastern Europe, respectively.

12. About 71%-73% of total cumulative investments in 1980-2000 would be electricity-associated investments. Total cumulative investment requirements for electricity and steam generation, transmission and distribution (shown in table 9) would be about 5,500 billion (1980 US) dollars in the low scenario and approximately 6,100 billion US dollars in the moderate scenario. More than half of total investments for the electricity systems would be for transmission and distribution.

13. By contrast, for the energy supply industry systems as a whole the share of investments for production would grow at the expense of transportation and distribution.

(d) Share of energy investments in total investments

14. In the past the "drain" of investments by the energy supply industries has been significant. The share of energy investments in total investments of these market economies absorbed an average of 15% with ranges from 4.0% (Denmark) to 23% (Canada, Norway) (see table 12). This share was generally increasing until the mid-1980s.

15. The share of investments 12/ in the fuel-energy complex of the USSR, in total investments industry-wide, steadily increased as well from more than 30% in 1980 to 40% in 1987 (see table 13). Nevertheless it should be outlined

that there exists some objective conditions which influence on investment requirements for the USSR energy supply industries such as bulk production of primary energy source, unique self-provision of the country with all energy sources, a large export of energy sources (300 mtoe in 1990), difficult geological and climatic conditions, etc.

16. The situation in east European countries (except the German Democratic Republic) followed an analogous trend; the share of investments in the fuel-energy complexes, in total industrial investments in these countries increased from approximately 30% in 1980 to 40% in 1986. This shows first, the importance of the energy branches in the economy of these countries and second, suggests that other investors with a high technological or scientific growth potential may have been affected by lack of funding.

D. DRIVING FORCES AND CONSTRAINTS

17. What are the forces that impact on the projected investment growth?

(a) **Driving forces**

18. Driving forces for energy investment requirements in the market economies and in eastern Europe have differed substantially until recent years. In connection with economic reforms market forces seem to play a more important part in the centrally planned economies in the future.

19. In the market economies major factors affecting the growth and structure of investments are changes in economic activity (change in output or capacity utilization), relative factor costs (real labour costs, real user costs of capital and real energy costs), profitability, impact of inflation, depreciation and tax laws on profits.

20. In eastern Europe the main determinants of investments are a shift from extensive to intensive development, efficiency of the investment process, foreign balance constraints, change in the technical and branch structure of investment, capital utilization and efficiency.

21. The size, structure and quality of indigenous energy deposits among other factors could be expected to act as a driving force affecting the volume of investments in both groups. Since widely varying conditions of extraction exist in the ECE region, the capital intensity of production and thus the investment made, should, to some extent, reflect these differences.

22. Slow-down in economic growth after 1973, the drop of profits, the rise of interest rates, fiscal constraints in the market economies and an increasing share of consumption with slower economic growth, deceleration in capital efficiency and strict foreign balance constraints in eastern Europe diminished the rate of growth of investment and investment ratios in these countries. This influenced in turn on expenditures in the energy supply sector.

23. The growth of population and labour supply influences on economic growth of each country and thus on investments including investments in energy supply industries. For example, population in the ECE region during 1985-2000 should rise at an annual growth rate of 0.7%, (medium variant), (North America 0.8%, Western Europe - 0.5%, eastern Europe - 0.7%). 13/

24. Current price levels encourage consumption and discourage the development of alternative resources or conservation efforts. However in some cases in eastern Europe they might stimulate energy conservation, diversification and internal substitution because the potential for energy conservation is large and has remained relatively underutilized.

25. But low oil prices cannot continue for ever. Some examples of energy branches suggest a reversal of the downward trends of capital-output ratios experienced in some countries. The following factors would exercise an upward pressure on the capital-output ratios.

26. Coal mining. Capital output ratios can be expected to continue slowly rising due to the exploitation of poorer deposits, environmental regulations, etc. in spite of an increasing share of strip-mining in some countries of the ECE region.

27. Oil and gas exploration. Greater investments per unit of output will be required due to the depletion of economically viable reserves, the development of marginal sources in severe climatic zones (Alaskan North Slope; North Sea; Siberia-Samotlor; Jamal peninsula; Arctic region, etc.), a growing share of off-shore oil and gas production and the growing application of secondary and tertiary extraction techniques. Scale and technical economies realized in new pipeline systems will be partly offset by increased distance of transportation. For example, consumption of gas for turbines of the giant gas pipelines from Jamal peninsula or Astrakan to other ECE countries would practically exceed exports and, according to some calculations, also exceed the net increment of natural gas production for the entirety of the USSR economy.

28. Electricity. Conventional power plants require additional investment (a) for environmental protection (an additional investment of 30-35% of capital costs); (b) for multifuel power stations at a time when the scope of scale economies appears exhausted; (c) hydropower stations whose capital-output ratios seem to increase due to environmental regulations and reduced potentialities; (d) for nuclear power whose capacities continue to increase substantially despite the incidents of Three Mile Island (United States) and Chernobyl (USSR). Various reasons (small territories, large distances, severe climatic conditions, etc) would require increasing capital-output ratios for electricity transportation, and distribution.

29. New and renewable sources of energy (solar, wind, geothermal, minihydro, shale oil, tar sands, etc.) are very capital-intensive.

30. An increase in the oil prices in 2010 by a factor of 2 (moderate scenario) could possibly promote energy savings. It has been shown in many cases that it is more economical to invest in energy conservation than to increase energy production capacity. 14/ For example, investment requirements for energy conservation in Hungary are 4-5 times lower than the investments in energy supplies. 15/

(b) Constraints

31. However, the growing share of investments in fuel-energy complexes competes with the investment requirements for heavy industry branches such as metallurgical and chemical industries. The problem is particularly pressing

in some east European economies which, up to now have imported a large share of their energy requirements mainly from the USSR. The USSR would either increase investments into engineering, light and food industries in order to improve standards of living and decelerate the share of investments in fuel energy complexes or continue making large expenditures for fuel energy complex. A continued emphasis on energy industries entails moving towards areas with severe climatic conditions, with poorly developed infrastructure, permafrost, swamps, etc. which require very large investments.

32. The latest decisions of the Soviet Government on accelerated growth of engineering branches spell out explicitly the urgent necessity to innovate and modernize obsolete equipment in industries including energy branches. The latest developments indicate a continuation of the previous growth of the share of energy branches in total investments, and a decelerating share of consumer goods industries (see table 13).

E. ALTERNATIVES

33. In the light of these demands, it is all the more important to evaluate whether the projected energy investment trends are sustainable or, even so, whether "cheaper" solutions can be found providing the same services.

34. A simple calculation may assist in approaching the first question. Energy supply investments in the ECE region are projected to grow at an average annual rate of 1 to 2%, depending on the economic growth scenario retained (para. 4 above). This would be significantly below the projected GDP rate of growth of 2.2-2.9%. For the market economies the estimates are 2.1-2.7%, for eastern Europe 2.4-3.5%. These relationships imply, ceteris paribus, that the share of energy supply investments in total investments is likely to fall through 2000, particularly in the market economies.

35. Still, this trend would not imply that the projected investment growth and pattern were "optimal". Indeed, it has been argued that there exist possibilities of achieving the same energy service with less investments.

36. One promising avenue, which however is controversial as to its potential, is investment in energy saving, energy efficiency, end-use oriented technologies and infrastructure. It has been suggested for example, 16/ that with best available and advanced technology, Sweden's per capita final energy use could be reduced by 2020 by 40 to 50% compared with 1975, depending on whether economic welfare increases by 50 or 100%. A United States country study along the same lines contends that based on the wide use of cost-effective energy-saving technologies by the year 2020 United States per capita energy consumption would fall by about 50%, while the consumption of goods and services would double.

37. In comparison with these projections, the increase of energy efficiency built in the energy demand/supply projections of the present study for the ECE countries while very significant are lower, particularly on a per capita basis.

38. Whatever the potential scope of a shift of investments from supplies to demand, and within supplies to less capital-intensive sources, the opportunities warrant a sustained co-operative effort.

Table 1

Penetration of Electricity in the ECE region economies in 1970-2010

Year	ECE region			North America & West.Europe			USSR and Eastern Europe		
	A	B	C	A	B	C	A	B	C
1970	3 877	965	25	2 794	707	25	1 083	258	24
1975	4 376	1 221	28	3 028	884	29	1 348	336	25
1980	4 931	1 457	30	3 350	1 029	31	1 582	428	27
1985	5 125	1 663	32	3 340	1 191	36	1 785	472	28
2000	6 100	2 350	39	3 850	1 700	44	2 300	670	29
2010	6 500	2 700	42	4 100	2 050	50	2 800	950	34

Source: ENERGY/AC.10/R.2, ECE Energy Data Bank: IEA Coal Information 1985-1988, Paris; Energetika SSSR v 1981-1985 godach, 1982 Moscow; Energetika SSSR v 1986-1990 godach, 1987 Moscow.

A. Gross primary energy consumption, million toe.

B. Inputs of energy sources to produce electricity, mtoe.

C. Inputs of energy sources to produce electricity in percentage over gross primary consumption, %.

Table 2

Changing fuel input for electricity generation in the ECE region
1980-2000
(Mtoe)

	Coal	Oil	Gas	Nuclear	Hydro, Geo-thermal	Others	Total
North America and Western Europe							
1980	462	125	107	114	218	3	1 029
1985	542	70	99	235	228	6	1 180
1990	582	74	96	314	253	7	1 324
2000	816	62	91	379	310	22	1 680
USSR and Eastern Europe							
1980	153	68	105	14	43	22	405
1985	136	67	137	47	54	31	472
1990	134	53	173	72	61	42	535
2000	152	40	185	171	86	56	690
ECE Total							
1980	615	193	212	128	248	25	1 441
1985	678	137	236	282	282	37	1 652
1990	716	128	269	386	314	49	1 859
2000	968	102	276	550	396	78	2 370
North America and Western Europe	%	%	%	%	%	%	%
1980	45	12	10	11	21	-	100
1985	46	6	8	20	19	1	100
1990	44	6	7	24	19	1	100
2000	49	4	5	23	18	1	100
USSR and Eastern Europe							
1980	38	17	26	3	11	5	100
1985	29	14	29	10	11	7	100
1990	25	10	32	14	11	8	100
2000	22	6	27	25	12	8	100
ECE Total							
1980	43	13	15	9	17	2	100
1985	41	8	14	17	17	2	100
1990	39	7	14	21	17	3	100
2000	41	4	12	23	17	3	100

Source: ENERGY/AC.10/R.2 ECE Energy Data Bank: IEA Coal Information 1985-1988, Paris; Energetika SSSR v 1981-1985 godach, 1982 Moscow; Energetika SSSR v 1986-1990 godach, 1987, Moscow.

Table 3

Electricity generation and generating capacities by energy sources in the market economies of the ECE region in 1980 and 2000

	1980		2000 low		2000 moderate	
Total primary energy consumption	3 350		3 701		4 000	
% used for electricity generation	31		45		42	
Total energy input in electricity generation, mtoe	1 029 c/	%	1 680	%	1 690	%
Total electricity generation, a/ TWh	4 116	100	6 720	100	6 720	100
– Hydro/geothermal/ others	888		1 328		1 328	
– Nuclear	456		1 516		1 516	
– Thermal	2 772		3 876		3 876	
Load factor b/						
– Hydro/geothermal/ others	0.42		0.42		0.42	
– Nuclear	0.55		0.66		0.65	
– Thermal	0.43		0.43		0.43	
Average	0.44		0.46		0.46	
Total electricity generating capacity, GWe	1 072	100	1 652	100	1656	100
– Hydro/geothermal/ others	241		361		361	
– Nuclear	95		262		266	
– Thermal	736		1 029		1 029	

Source: ENERGY/AC.10/R.2 ECE Energy Data Bank: IEA Coal Information 1985-1988, Paris; Energetika SSSR v 1981 - 1985 godach, 1982 Moscow; Energetika SSSR v 1986-1990 godach, 1987, Moscow.

a/ 1 TWh requires 0.25 mtoe (efficiency of about 34%).

b/ Load factor = energy production divided by potential production.

c/ Calculated from total electricity generation assuming that 1 TWh requires 0.25 mtoe.

Table 4

Electricity generation and generating capacities by energy sources in
central and east European countries of the ECE region in 1980 and 2000

	1980		2000 low		2000 moderate	
Total primary energy consumption, mtoe	1 582		2 216		2 335	
% used for electricity production	27		29		31	
Total energy input in electricity generation, mtoe	428 c/	%	654	%	690	%
Total electricity generation, a/ TWh	1 713	100	3 090	100	3 270	100
– Hydro/geothermal/ others	210		340		380	
– Nuclear	83		540		580	
– Thermal	1 421		2 210		2 320	
Load factor b/						
– Hydro/geothermal/ others	0.38		0.40		0.40	
– Nuclear	0.54		0.66		0.65	
– Thermal	0.59		0.58		0.58	
Average	0.55		0.56		0.56	
Total electricity generating capacity, GWe	356	100	630	100	667	100
– Hydro/geothermal/ others	63		98		108	
– Nuclear	18		93		102	
– Thermal	276		439		457	

Source: ENERGY/AC.10/R.2 ECE Energy Data Bank: IEA Coal Information
1985-1988, Paris; Energetika SSSR v 1981-1985 godach, 1982, Moscow;
Energetika SSSR v 1986-1990 godach, 1987, Moscow.

a/ 1 TWh requires 0.25 mtoe (efficiency of about 34%).

b/ Load factor = energy production divided by potential production.

c/ Calculated from total electricity generation assuming that 1 TWh
requires 0.25 mtoe.

Table 5

Electricity generation and generating capacities by energy sources in the ECE region in 1980 and 2000

	1980		2000 low		2000 moderate	
Total primary energy consumption	4 931		5 917		6 335	
% used for electricity generation	30		38		38	
Total energy input in electricity generation, mtoe	1 457 c/	%	2 334	%	2 370	%
Total electricity generation, a/ TWh	5 829	100	9 332	100	9 480	100
- Hydro/geothermal/ others	1 098	19	1 672	18	1 706	18
- Nuclear	539	9	2 052	22	2 096	22
- Thermal power stat.	4 193	72	5 608	60	5 678	60
Load factor b/						
- Hydro/geothermal/ others	0.41		0.42		0.42	
- Nuclear	0.55		0.66		0.65	
- Thermal	0.47		0.44		0.44	
Average	0.47		0.47		0.47	
Total electricity generating capacity, GWe	1 429	100	2 282	100	2 323	100
- Hydro/geothermal/ others	304	21	459	20	469	20
- Nuclear	113	8	355	16	368	16
- Thermal	1 012	71	1 468	64	1 486	64

Source: ENERGY/AC.10/R.2 ECE Energy Data Bank: IEA Coal Information 1985-1988, Paris; Energetika SSSR v 1981-1985 godach, 1982, Moscow; Energetika SSSR v 1986-1990 godach, 1987, Moscow.

a/ 1 TWh requires 0.25 mtoe (efficiency of about 34%).

b/ Load factor = energy production divided by potential production.

c/ Calculated from total electricity generation assuming that 1 TWh requires 0.25 mtoe.

Table 6

Development of electric generating capacity 1980 and 2000, GWe

	Total electric generating capacity			Cumulative replacement capacity (retirements)	Cumulative net additions to capacity		Cumulative total capacity additions	
	1980	2000 low	2000 moderate	1980 - 2000 low-moderate	1980 - 2000 low-moderate		1980 - 2000 low-moderate	
ECE								
Total	1 429	2 282	2 323	927	853	895	1 780	1 822
-Hydro/geothermal/ others	304	459	469	115	155	165	270	280
-Nuclear	113	355	368	85	242	255	327	340
-Thermal	1 012	1 468	1 488	727	456	474	1 183	1 201
North America and Western Europe a/								
Total	1 072	1 652	1 656	761	580	584	1 341	1 345
-Hydro/geothermal/ others	241	361	361	96	120	120	216	216
-Nuclear	95	262	266	76	167	171	243	277
-Thermal	736	1 029	1 029	589	293	293	882	882
USSR and Eastern Europe b/								
Total	357	630	667	166	273	311	439	477
-Hydro/geothermal/ others	63	98	108	19	35	45	54	64
-Nuclear	18	93	102	9	75	84	84	93
-Thermal	276	439	457	138	163	181	301	319

Source: ENERGY/AC.10/R.2 ECE Energy Data Bank: IEA Coal Information 1985-1988, Paris;
Energetika SSSR v 1981-1985 godach, 1982, Moscow; Energetika SSSR v 1986-1990 godach, 1987,
Moscow. H.K. Schneider, W. Schulz: "Investment Requirements of the World Energy Industries
1980-2000", WEC, 1987.

a/   Retirements of 40% of existing hydro/geothermal/other capacity in 1980. 80% of existing
nuclear/conventional thermal capacity in 1980.

b/   Net additions to capacity 1980-2000 = capacity in 2000 minus capacity in 1980.

c/   Retirements of 30% of existing hydro/geothermal/other capacity in 1980. 50% of existing
nuclear/conventional thermal capacity in 1980. In Energy Programme of the USSR the total
replacement capacity of power stations was estimated to 150 GWe eg. 50% of total installed
capacities.

Table 7

Per unit investment requirements for electricity generation, transportation and distribution, 1980 and 2000 by regions
US - $ (1980) kWe

	1980	2000 low scenario	2000 high scenario
Electricity generation			
By sources:			
- Hydro/geothermal/ others	2 000	2 450 a/	3 000 b/
- Nuclear	1 500	1 850 a/	2 250 b/
- Conventional thermal	750	900 a/	1 100 b/
By regions: c/			
- Industrialized countries (IC)	1 080	1 460	1 740
- Developing countries (DC)	1 230	1 510	1 810
- Countries with planned economies(CPE)	1 000	1 320	1 690
Electricity transmission & distribution **By regions:**			
- Industrialized countries (IC)	2 000	2 020	2 210
- Developing countries (DC)	590	1 240	1 810
- Countries with planned economies(CPE)	1 000	1 430	1 910
Breakdown of total per unit investment requirements, % **IC's**			
- generation	35	42	44
- transmission & distribution (T&D)	65	58	56
DC's			
- generation	66	55	50
- transmission & distribution (T&D)	34	45	50
CPE's			
- generation	50	48	47
- transmission & distribution (T&D)	50	52	53

Source: H.K. Schneider, W. Schulz: "Investment Requirements of the World Energy Industries 1980-2000", WEC, 1987.

a/ 1%/yr average real rate of increase over the period 1980-2000 values in the table rounded off.

b/ 2%/yr average real rate of increase over the period 1980-2000 except nuclear power, values in the table rounded off.

c/ Weighted averages using the capacity shares from Tables 6.1 - 6.3.

Table 8

Investment requirements for electricity generation from renewable and non-commercial energy sources

	Per unit investment requirements US $ (1980) kWe	
	1980	2000
Geothermal	2 000	2 500
Mini-hydro	2 500	3 500
Wind	3 500	2 500
Solar thermal energy system (25 kWe module)	4 000	2 000
Photovoltaics	10 000 – 15 000	1 400 – 1 900
Peat	1 080	1 520

Source: Oil Substitution World Outlook to 2020, WEC, 1985 pp. 273, 274, 276; New and Renewable Energy Sources, OECD, 1987, Paris, pp. 145, 170.

Table 9

Cumulative investment requirements for electricity generation

	Per unit investment requirement US $ - (1980)/kWe			Cumulative capacity additions GWe		Cumulative investment requirements Billion US - $ (1980)	
	1980	2000 low	2000 moderate	1980 - 2000 low	- moderate	1980 - 2000 low	- moderate
ECE							
- generation				1 780	1 822	2 222	2 542
of which:							
coal						948	984
oil						169	195
- transmission & distribution						3 233	3 536
- Total						5 455	6 078
North America & Western Europe							
- generation	1 086	1 460	1 740	1 341	1 345	1 708	1 896
of which:							
coal						806	805
oil						118	131
- transmission & distribution						125	138
distribution	2 000	2 020	2 210			2 695	2 837
- Total						4 403	4 733
Eastern Europe, USSR							
- generation	1 020	1 320	1 690	439	477	514	646
of which:							
coal						142	179
oil						51	64
- transmission & distribution						136	171
distribution	1 020	1 430	1 910			538	699
- Total						1 052	1 345

Source: ENERGY/AC.10/R.2 ECE Energy Data Bank: IEA Coal Information 1985-1988, Paris; Energetika SSSR v 1981 - 1985 godach, 1982, Moscow; Energetika SSSR v 1986-1990 godach, 1987, Moscow H.K. Schneider, W. Schulz: "Investment Requirements of the World Energy Industries 1980-2000", WEC, 1987.

Table 10

**Absolute investment requirements of the ECE energy supply industries
1980 and 2000**

	Oil			Natural Gas			Coal			Electricity			Total		
	1980	2000 low	2000 mod.	1980	2000 low	2000 mod.	1980	2000 low	2000 mod.	1980	2000 low	2000 mod.	1980	2000 low	2000 mod.
ECE	50	57	76	17	21	32	38	37	51	192	246	283	297	361	442
Market econom.	31	30	40	16	18	29	31	30	40	148	184	193	226	262	302
Eastern Europe, USSR	19	27	36	1	3	3	7	7	11	44	62	90	71	99	140

Source: H.K. Schneider, W. Schulz: "Investment Requirements of the World Energy Industries
1980-2000", WEC, 1987.

Table 11

Total cumulative and per unit energy investment requirements of ECE
1980 and 2000 by energy sources and functions

Billion US - $ (1980), US - $ (1980)/toe

	1980-2000 low scenario							1980-2000 moderate scenario						
	Oil	Natural Gas	Coal	New Energy Sources a/	Non-Commercial a/	Electricity	Total	Oil	Natural Gas	Coal	New Energy Sources a/	Non-Commercial a/	Electricity	Total
ECE countries Cumulative production, 10⁹toe	28.0	27.0	32.0	0.7	3.0	17.0	107.7	29.0	28.0	33.0	1.0	3.0	18.0	112.0
Market economies " "	15.0	14.0	18.0	0.5	2.0	12.0	61.5	16.0	14.0	19.0	0.7	2.0	12.0	63.7
Eastern Europe, " " USSR	13.0	13.0	14.0	0.2	1.0	5.0	46.2	13.0	14.0	14.0	0.3	1.0	6.0	48.3
Cumulat.invest.10⁹ ECE, US - $ (1980)"	598.0	837.0	320.0	35.0 b/	18.0c/	5 455	7 264	1 073.0	1 120.0	528.0	50.0 b/	20.0 c/	6 078.0	8 869.0
Market economies " "	329.0	569.0	179.0	25.0	18.0	4 403.0	5 523.0	601.0	784.0	333.0	35.0	19.0	4 733.0	6 505.0
Eastern Europe, " " USSR	269.0	268.0	141.0	10.0	1.0a/	1 052.0	1 741.0	472.0	336.0	195.0	15.0	1.0	1 345.0	2 364.0
Per unit investment US - $ (1980)/toe	34.0	31.0	10.0	500.0	15.0	321.0	71.0	37.0	40.0	16.0	500.0	15.0	338.0	79.0
Production	25.0	14.0	8.0	500.0	15.0	144.0	37.0	28.0	16.0	12.0	500.0	15.0	152.0	41.0
Transport & distri.	9.0	17.0	2.0	0.0	0.0	177.0	34.0	9.0	24.0	4.0	0.0	0.0	186.0	38.0

Source: ENERGY/AC.10/R.2 ECE Energy Data Bank: IEA Coal Information
1985-1988, Paris, Energetika SSSR v 1981-1985 godach,
1982, Moscow, Energetika SSSR v 1986-1990 godach, 1987, Moscow
H.K. Schneider, W. Schulz: "Investment Requirements of the World Energy Industries 1980-2000", WEC, 1987.

a/ Mainly utilized in electricity production.

b/ For heat requirements.

c/ Used for biogas production, grain drying, cooking, water heating, etc.

Table 12

Percentage share of investments in energy supplies, in total fixed industrial investments

Belgium		
	1965-1973	7.3
	1973-1980	7.6
	1980-1982	7.3
Canada		
	1965-1973	15.2
	1973-1980	17.8
	1980-1982	22.9
Denmark		
	1965-1973	4.7
	1973-1980	4.1
	1980-1982	...
Finland		
	1965-1973	6.1
	1973-1980	8.8
	1980-1982	5.9
France		
	1965-1973	7.7
	1973-1980	7.8
	1980-1982	...
Germany, Federal Rep. of		
	1965-1973	6.1
	1973-1980	7.0
	1980-1982	6.2
Ireland		
	1965-1973	5.8
	1973-1980	6.1
	1980-1982	8.2
Italy		
	1965-1973	9.3
	1973-1980	9.6
	1980-1982	11.2
Netherlands		
	1965-1973	9.4
	1973-1980	8.0
	1980-1982	7.8
Norway		
	1965-1973	10.3
	1973-1980	21.8
	1980-1982	23.2
Sweden		
	1965-1973	9.1
	1973-1980	9.9
	1980-1982	10.4
United Kingdom		
	1965-1973	12.5
	1973-1980	14.8
	1980-1982	18.1
United States of America		
	1965-1973	14.6
	1973-1980	14.4
	1980-1982	13.5

Source: Economic Bulletin for Europe, Vol. 38, No. 2, p. 254

Table 13

Share of investments of some branches in the industry of the USSR, %

Branch	1971-1975 IX Five-Year Plan	1976-1978 X Five-Year Plan	1981-1985 X1 Five-Year Plan	1986	1987
Industry	100.0	100.0	100.0	100.0	100.0
Fuel-Energy complex	29.1	30.0	36.3	38.6	40.1
Light industry	4.1	3.8	3.7	3.2	2.7
Food industry	7.1	6.5	6.1	5.7	5.5
Other industry 1/	59.7	59.7	53.9	53.9	51.7

Source: The industry of the USSR; Statistical Abstract of the USSR (Russian) Moscow, 1988, pp. 53, 78.

1/ Engineering, iron and steel, chemical, non-ferrous metals industries, etc.

ENERGY TECHNOLOGIES FOR THE FIRST DECADES
OF THE TWENTY-FIRST CENTURY 17/

I. Introduction

The first decades of the twenty-first century are roughly 50 years away and one might consider this too long a time distance for any meaningful assessment of technological matters. There are indeed technical areas where such a statement is most appropriate, micro-electronics is certainly a case in point. But for the case of energy technologies it is rather the opposite, their dynamics is very slow and long range in nature. This has rather important repercussions.

On the one hand there are the market forces of the day. They are price driven, the importance of the oil price demonstrates that clearly. It leads the energy market and thereby related technologies. Any such technological development must stand the proof of competitiveness. Measures for energy conservation at the user's end are an example.

On the other hand there are the technological forces that change the structure of the market and drive thereby the prices. They in turn are driven by science and society. C. Marchetti has studied for many years the dynamics of these technological forces. A case in point is the substitution of primary energy carriers. Figure 1, which is widely known today, exhibits the change from wood to coal to oil to gas. It shows the respective relative market shares in the particular plot where a S shape logistic transition appears as a straight line. One remarkable thing of Figure 1 is the fact that it covers a time span of 120 years and more. And it is remarkable that there exist regularities over such long periods irrespective of world wars or economic crises.

An expertise on energy technologies for the first decades of the twenty-first century must take into account both partly conflicting aspects, the short-range and the long-range time horizons.

Forecasts are feasible under short range time horizons, say a few years, perhaps up to five. A principal tool for such forecasts are the well established econometrical methods. By identifying governing parameters in particular elasticities from time series that characterize a market it is possible to forecast reactions of such markets once the development of prices or incomes or other such driving variables are known.

Long range time horizons do not allow for forecasts in the immediate sense of the word. It is not possible to predict the one future that is going to become reality in the decades ahead. What is possible is to anticipate possible futures, the plural. The principal tool for such anticipation is scenario writing. It conceives in a consistent way comprehensive pictures of possible futures. This includes population and economic growth, the availability of resources, the introduction of new technologies and other features. Assumptions must be made explicit and consistent. In view of the nature of the here considered problem a reference to such scenarios is mandatory. Because the elaboration of a scenario is a lengthy and tedious task no attempt is made to elaborate a new one. Instead, a few existing scenarios are being reviewed and later used appropriately, especially in view of demand projections.

II. A review of existing energy scenarios

A. The IIASA-Low-Scenario

Two globally comprehensive scenarios were produced by the Energy Systems Program Group of the International Institute for Applied Systems Analysis (IIASA) as early as 1975 and published in 1981 [1]. The "High" Scenario follows the idea of a growth of the Gross Domestic Product (GDP) in the Developing Countries (DCs) close to 4-5%, a figure that was much demanded in the debates of the "New Economical Order" in the early seventies [2]. In order to achieve such a relatively high growth rate there it was necessary to assume growth rates close to 3% in the industrialized countries (ICs) [3] thus arriving at a world's total primary energy demand of 35 TW years/year in 2030. For purposes of comparison: In 1975 when the IIASA Scenarios were conceived the world consumption of primary energy was 8 TW years/year. In 1988/89 it is close to 11 TW years/year. Today, an expectation of such an increase of primary energy consumption is not very fashionable, times have changed but will continue to change.

The IIASA "Low" Scenario assumes lower GDP growth rates. For the DCs 3% is a typical figure and for the ICs 2%. The so expected consumption of primary energy is accordingly lower, 22 TW years/year. Since 1975, the base year of the two IIASA Scenarios, 14 years have elapsed and it is therefore interesting to make a comparison with the actual development since that time. To that end it is important to realize the seven IIASA World Regions as explained in Figure 2. These Regions are not necessarily geographically coherent. Instead, they reflect features of economy, population growth and resources which they have, to a degree, in common. The two IIASA Scenarios are geographically disaggregated accordingly. It turns out that the IIASA Low Scenario is close to the actual development since 1975. Table 1 compares the IIASA Low data with the actual development as given by the BP statistics for the seven World Regions. Given the fact that the aggregation of the BP data into the IIASA Regions poses some difficulties it turns out that the IIASA Low Scenario is rather close to the actual development over the years between 1975 and 1987, not only for the world as a whole but also, to varying degrees, for the World Regions individually.

The IIASA Low Scenario conceives - as already observed - a 22.39 TW years/year level of primary energy supply for 2030 thus essentially reflecting a 2% overall growth rate of primary energy for the period of 1988 to 2030. This is not an unreasonable figure. Reasonableness refers to growth rates in the individual Regions, a degree of energy conservation, and to changes in the structure of the economy [4]. It thus may provide a yardstick for further consideration in this paper. For the purposes of this paper it is important to concentrate on the countries that are members of the Economic Commission for Europe (ECE). The IIASA World Regions nicely permit for that:

Region I	comprises North America, that is the United States of America and Canada
Region II	comprises the planned economies of Eastern Europe including all of the Soviet Union
Region III	comprises the OECD countries other than North America

Table 2 relates the individual countries to these three IIASA
World Regions. While indeed the Regions I, II and III are here in the
forefront of interest it is nevertheless important to keep in mind that we
live in a Finite World and any energy strategy interrelates to all others.
Therefore it is useful to use scenarios for the regions of ECE that are part
of world-wide comprehensive scenarios.

B. The WEC-Scenarios of J.R. Frisch

The WEC also produced two scenarios that are of interest here:
M and L [5], J.R. Frisch being the Project Director. They cover the years
from 1960 through 2020 for five regions that are world-wide comprehensive.
This poses no real problem here when the ECE countries are considered. In
fact, the following relations are valid:

WEC, North 1 <-----> IIASA, Region I + III

WEC, North 2 <-----> IIASA, Region II

The allocation of the individual countries of the WEC regions of interest
here is given in Table 3.

It turns out that the GDP data of the IIASA Low Scenario are somehow
close to the WEC, M data (except for North 2, 2030), Table 4. The primary
energy data of the IIASA Low Scenario are mostly somewhat higher than the WEC,
M data. It may be worth noting that the overall energy intensity is
consistently not far from 1 Watt year/$ in North 2, that is the Soviet Union
and the Eastern European Countries while for North 1, that is the OECD, such
intensities are close to 0.5 Watt year/$. Table 5 gives more recent
information on this important characteristic. The problem of regional
comparisons that involve GDP relates mostly to problems of inflation, exchange
rates and changes of economical structures. But in any event one recognizes
the widespread that reaches from Japan with 0.22 W year/$ 87 to the People's
Republic of China with 2.66 W year/$ 87, a factor of 12! For the purposes of
this paper we will make use of the IIASA Low Scenario for reference purposes.

C. The Energy Conservation Scenarios of U. Colombo/Bernardini and
    J. Goldemberg et al.

The most hotly debated issue today is energy conservation by improving
energy end use efficiencies. It is necessary to present related information
here and to relate it also to certain political developments.

There is a fundamental statement of those who want to introduce
significant energy conservation measures: One can enjoy the same services
from energy end uses like a warm room or an accomplished transportation by
introducing technologies with higher efficiencies, primarily on the users' end
but also including the conversion from primary to secondary energy. What is
not so often and explicitly stated is the implication that this means in most
cases a substitution of energy by capital as well as labour and skill.
Indeed, an old steam locomotive used about five times as much energy as a
modern Diesel locomotive for the same transportation service. But a Diesel
locomotive requires more capital and more skill. And one can isolate a room
further and further by the investment of more and more appropriate materials
and thereby decrease the energy for heating, but this requires capital and

skill. Therefore the energy issue is transformed into a much broader economic and society related issue and one has to apply extreme caution and care before drawing far reaching conclusions. This is particularly so when not only marginal degrees of energy conservation measures are introduced but substantial amounts. Marginal amounts appear as easy and cheap [6] but this is generally misleading. The characteristic that is used for the introduction of significant degrees of energy conservation is the above-mentioned energy intensity, the ratio between energy used and GDP produced.

U. Colombo and O. Bernardini have pronounced energy conservation in a scenario that was conceived in 1979 [7]. Their basis was to keep the then prevailing 2 kW years/year cap of energy consumption per average global inhabitant constant as the world population increases and the various economies of the DCs develop. Any growth of related energy consumption in the DCs must therefore be compensated by a decrease of energy intensities in the ICs and a corresponding conservation of energy. Should the economies of the ICs experience a growth of their own such conservation of energy must go even further. It should be stated here explicitly that in this report the related impacts on the economies and societies in question cannot be evaluated or foreseen. That must remain open here.

When in 1975 the world population was about 4 billion people a related 2 kW years/year cap yielded a world's total consumption of 8 TW years/year. Then a population of 8 billion people in 2030 yields 16 TW years/year in the Colombo/Bernardini scenario. That scenario uses the seven IIASA World Regions and therefore it was possible to compare the 16 TW year/year case with the IIASA Low Scenario still in the principal IIASA publication [1]. Figure 3 and Figure 4 give the implied energy intensities. Absolute numbers have to be dealt with care because of the above-mentioned monetary uncertainties but what matters are comparisons and trends. The intensity ratios of the IIASA Low case exhibit a somehow consistent trend when historical data are extended to scenario data. One should note that the base year was 1975. By contrast, the Colombo/Bernardini case shows sharp reversals of trends. Without such sharp reversals a 16 TW year/year target for 2030 cannot be achieved. Table 6 illustrates the problem further. There the data for final energy (e.g. electricity at the plug or gasoline at the filling station) are being considered for Regions I and III which make up for North 1 of the WEC scenarios. One realizes that by 2030 the Colombo Scenario assumes essentially the same figures as for 1975, a zero growth. Or in other words, any growth of energy services must come from energy conservation in these regions, while the IIASA Low Scenario envisages such a growth. The degrees of conservation thus vary between 8 and 54%. The Goldemberg et. al. scenario goes even further [8]. It assumes not only a constant per capita value but an overall constant consumption of primary energy, namely that of 1987, and that is 11 TW years/year. A growth of population and a growth of economies thereby result in even greater requests for energy conservation.

Table 7 and Table 8 thus compare both conservation scenarios with the IIASA Low case. It should be noted that the values for the DCs in 2030 do not differ greatly in the three scenarios. What matters is the degree of the expected energy conservation in the industrialized countries, that is the Regions I + II + III. Against the IIASA Low case Colombo/Bernardini imply an energy conservation of 43% of primary energy for the year 2030. In the case of Goldemberg the number is 69% for the year 2020.

One should note the fact that the Goldemberg approach as outlined in a recent study of him and his colleagues [9] had a determining influence on the widely recognized so-called Brundtland Report [10]. This Brundtland Report is named after the Norwegian Prime Minister Gro Harlem Brundtland who served as the chairperson of the World Commission on Environment and Development, Our Common Future.

III. Major constraints

The IIASA Low Scenario, the Colombo/Bernardini as well as the Goldemberg Scenario were meant to be general scenarios. Above two constraints were identified, namely global warming and depletion of reserves respectively resources. Whatever the differences of the three general scenarios were the following 1989 Jülich Scenario is meant to respond specifically to the constraint of global warming.

A. Global warming

At the latest since the Toronto Conference of 1988 [11] a wide public is rightfully paying attention to the future of the global atmosphere. It is not the purpose of this paper to elaborate on that. Therefore a few remarks must suffice.

The pre-industrial level of the CO_2 concentration in the atmosphere was close to 260 ppm.v. Presently it has increased to 340 ppm v and increases further. If the present trends continue a doubling is expected by the middle of the next century. Besides of the emissions of CO_2, mostly upon burning of fossil fuels, there are additional emissions that essentially double the greenhouse effect. Such additional emissions come from N_2O, methane and the CFCs (chlorofluorocarbons). A doubling of the effective CO_2 content - and that includes the equivalent of the other greenhouse gases - is expected to increase the global average temperature by 3 ± 1.5 K. But such average temperature increase is unevenly distributed across the geographical latitudes. At the poles it could be as much as 10 K causing thus the melting of ice shelves and subsequently the increase of the sea level by roughly one meter. Of equal or even larger significance may be the expectation of a preceding enhanced climate variability. Some people relate the climate anomalies that we presently experience already to that enhanced variability. A share of about 50% of such CO_2 emissions does not stay airborne but is accepted by the upper layer of the oceans. There it takes a period of 500-1,000 years until such CO_2 content is given to the bottom of the sea, the final waste disposal site for such CO_2. The dynamics of the oceans provide for such mechanism. But such mechanisms are not yet sufficiently understood. It may be possible that they provide for the opportunity of a fixed ration of CO_2 that is sustainably given to the deep sea. The amount of such CO_2 ratio is not yet known, but it is estimated to be perhaps at 2.5 GtC/year (Gigaton of Carbon in the chemical form of CO_2 per year) [12]. Eventually a future world's energy system should live up to such a ration as a limit for using the carbon atom. Presently the energy related emissions are at 6 GtC/year and 1 GtC/year must be added coming from non-energy related sources. The Toronto Conference has asked for an overall reduction of 20% of energy related carbon emissions, that is down to 4.8 GtC/year, by 2005.

It is along such lines that the 1989 Jülich CO_2 reduction scenario aims for an emission of only 4 GtC/year by 2030 with the understanding that further reductions are necessary and possible during the rest of the next century. A transition to energy systems that mainly use electricity and hydrogen on the secondary energy side and nuclear and/or solar energy on the primary energy side make that principally possible.

There exist six possibilities for the reduction of CO_2 emission:

(a) a change of the mix of fossil fuels towards a high hydrogen content;

(b) the engagement of biomass as a fuel which is considered to permit for the recycling of the carbon atom;

(c) the engagement of carbon free solar based energy sources such as hydropower, wind and direct uses of solar power;

(d) the large-scale engagement of nuclear power;

(e) the engagement of energy conservation, primarily through improvement of efficiencies;

(f) CO_2-recovery - disposal techniques.

Having dealt with the element of energy conservation respectively efficiency improvements of energy end uses one also has to evaluate the other possible measures for the reduction of CO_2 emissions.

ad. a.)
A change of the fossil fuel mix can possibly help significantly. Table 9 gives the carbon content in fossil fuels as an estimated global average. For one unit of energy natural gas releases only 57%, and oil 80% of the CO_2 amounts that are released by coal. But one has to realize that methane is a greenhouse gas itself and contributes to global warming 32 times as much as CO_2 when compared on a molecule by molecule basis. The effective carbon content equivalent therefore follows the relationship

$$\frac{GtC}{TWa} = 0.432 \ (1 + 1 \times 32)$$

if 1 is the ratio of methane moles lost per methane moles actually oxidized. If 1 equals 2.3% the effective carbon content of methane equals that of coal. Actual losses of methane that remain in the atmosphere are in that range but not really known yet. For the purposes here we therefore assume that 1 can be kept small but attention must be paid to further related information. C. Marchetti has proposed [13] to reform natural gas by the use of the High Temperature Reactor (HTR) according to the reaction

$$CH_4 + 2 \ H_2O \ ----> CO_2 + 4 \ H_2$$

His idea is to pump the CO_2 so generated into the very same underground caves where the CH_4 was taken from. This would be a novel use of methane together with nuclear power that does not generate greenhouse gases. The technology of applying exogeneous heat to the reformation of methane has been successfully proven at KFA Jülich in the 10 MW range. It is known as the EVA scheme [4].

ad b.)

The use of biomass follows the idea of carbon recycling. Therefore if one unit of biomass is burned per unit of time one such unit of biomass must be replanted per unit of time. Goldemberg et al. envisage in principle a global potential of as much as 5 TW years/year [15]. As such this is a sound idea. Nevertheless, one should be aware of its implications. Under natural circumstances the power production density of biomass is close to 0.2 W/m^2 [16]. Assuming an enhanced value of 0.3 W/m^2 a total of 5 TW years/year would imply an area of about 17 mio km^2. The present global agricultural uses of land make up for about 13 mio km^2. Besides the sheer size of such additional land uses such large scale uses of biomass imply large logistic problems of transport and management. One should bear in mind that the average power consumption density in urban areas is 5 W/m^2 [17]. A city like Vienna covers an area of about 600 km^2 for its supply on the basis of biomass it would therefore require 10,000 km^2. Given streets, villages and other uses of land this would come effectively close to an area of perhaps 150 km x 150 km. Indeed, prior to the first industrialized revolution when wood was the principal source of energy such land use patterns existed around Vienna and created severe problems. There will be other problems like ecological consequences or recycling of nutrients. The exact nature of these implications is certainly not fully understood yet. They vary also from region to region and are certainly different in Brazil and Austria. One therefore can only assume that a total of perhaps 1.5 TW years/year or so is a more realistic guess. And such a value was indeed used in the Goldemberg scenario.

ad c.)

Carbon free solar power refers to hydropower, to local uses of solar power and to large scale central uses of solar power. The global total of hydropower is somewhere at 1 TWe years/year and therefore of the same order as the above considered case of biomass. Much of such potential of hydropower is already used. This is notwithstanding the fact that locally additional uses of hydropower lend themselves very much. Places in South America and Africa provide examples for that. Large scale uses of solar power provide indeed the opportunity for significant supplies of power in the several TW range. In sunny arid areas, say the north African deserts, a value of 10 We/m^2 gives a reasonable overall average. At 1 mio km^2, in these desert areas not an unreasonable figure, this results in as much as 10 TWe. It is generally known that the overall costs are still prohibitive. Only partly this relates to the costs of photovoltaics, much goes to power conditioning and other infrastructures. It is not the purpose of this paper to elaborate on future uses of solar power and other alternative sources, it is nuclear power that is of interest here. Therefore the above remarks should suffice.

It turns out that all of the possibilities must be engaged should there be any hope for the achieving of the 4 GtC/year goal. The idea of the 1989 Jülich CO_2 reduction scenario is to go to the limits for the possibilities a.), b.), c.) and e.) and then to give the remainder to nuclear energy, that is d.). Accordingly it is now that the scenario must be introduced, thereafter will then the degree of using nuclear power be understood.

It is now possible to conceive a CO_2 reduction scenario against the background of the considerations made above. This will in turn be instrumental for the assessment of the role of future uses of nuclear power in such a context. A number of explanations are in place:

(i)   The scenario assumes the reality of the CO_2 effect.

(ii)   The scenario is meant for 2030. By then an ultimately satisfactory solution of the CO_2 problem can under no circumstances be expected. Therefore the scenario can only describe a transitional stage of world's energy systems.

(iii) Given the situation described under (ii) the scenario aims at a value of 4 GtC/year emission rate of CO_2. With 1 GtC/year coming from non-energy sources this results in the 5 GtC/year mentioned above. Should they not continue at that scale the CO_2 concentration would be accordingly less. Compared with the 1987 energy related emissions of 6 GtC/year this would be a reduction of 33% of that value.

(iv)   The scenario is not the result of a mathematical model of any kind. But it does observe a number of constraints and follows certain reasons.

       Within the space so left it is judgemental and thus it asks for further work.

(v)   The scenario is not meant to be feasible in todays meaning of the word. Todays feasibilities do not allow for significant CO_2 reductions. Instead, the scenario exhibits a number of "impossible" features and essentially addresses the question: How impossible is impossible?

Given these explanations it is fundamental to assume the degree of energy conservation. In that respect the 1989 Jülich CO_2 reduction scenario follows the scenario of Colombo and Bernardini thus assuming a degree of energy conservation in the industrialized part of the world that is close to 43% of the primary energy demand as foreseen by the IIASA Low Scenario. This is felt to be on the high side of energy conservation. To what extent that is "possible" or "impossible" must remain open here.

The acceptance of features of the Colombo/Bernardini scenario goes further. This scenario makes use of the seven IIASA World Regions and when foreseeing 16 TW years/year for the year 2030 it also allocates shares of that primary energy to the IIASA World Regions. Also these allocations are accepted. The question is then left how these shares of 16 TW years/year can be provided by the uses of the various primary energy sources in such a way that a total of 4 GtC/year is not superceded. The answer to that question is in fact more difficult than it appears at a first glance.

Table 10 presents the 1989 Jülich CO_2 reduction scenario.

The allocation of oil is straightforwardly idealistic. Oil is the most precious form of primary energy requiring only a minimum of capital intensive infrastructure. Thus a major share of the overall figure of 3.5 TW years/year is given the DCs while the total of 1987 was 4.19 TW years/year. For the OECD countries (IIASA-Regions I + III) this would mean a halving of its present consumption, for China, Africa and South East Asia essentially a doubling.

The traditional uses of natural gas mostly continue. Certain reductions are expected for the Soviet Union while a doubling is expected for Latin America. 1987's total of 2.2 TW years/year is reduced to 2.0 TW years/year. But there are the new uses of natural gases (gas_2) together with nuclear power ($nuclear_2$) as described above, resulting in hydrogen. Such hydrogen may be transportable and such imports are assumed for Western Europe, Japan, Latin America and Africa (IIASA-Regions III, IV and V) where natural gas sources are limited. But it is left open whether hydrogen or methane is transported. One should realize that such novel uses double the uses of natural gas, another 2 TW years/year are engaged, but together with 1.5 TW years/year of high temperature nuclear process heat. Otherwise it is not possible to meet the target of 4 GtC/year.

Coal in turn must be reduced by more than half, from 1987's 3.39 TW years/year down to 1.5 TW years/year. This is a serious cut for North America, more so for the Soviet Union and for Western Europe. But the most serious case is that of China which has to decrease somewhat its 1987 value instead of increasing it to 3 TW years/year what is presently envisaged by State planners in that country.

Solar with a direct electrical output (wind, Photovoltaics) are given a total of 1.2 TW years/year. That is more than twice the nuclear energy generation of 1987. It is purposely a high figure. The perception is that it could be less but not more. Most of these 1.2 TW years/year are seen in the Middle East and Africa (Region V). All the allocations to the Regions as well as the overall figure are debatable, but resulting probably in lower values.

Hydro and Bio Sources are somewhat in line with estimates of the Goldemberg Scenario. This leaves 2.2 TW years (thermal)/year for nuclear electricity generation. It is on purpose that most of it is meant to come from the industrialized part of the world (Regions I-III). But China is given also a large share, otherwise it could not compensate for the waiver of additional coal uses.

The result of such a disaggregated approach is given in Table 11.

ad d.)

The 1989 Jülich CO_2 reduction scenario yields 4.11 GtC/year. By
comparison, the Colombo/Bernardini scenario yields 5.18 GtC/year and
Goldemberg 4.79 GtC/year.

The following helps to understand the assessed role of nuclear power:
For reasons of comparison one may want to reduce the role of nuclear power in
the 1989 Jülich reduction scenario down to a value of 0.75 TW years/year, the
value that was considered by Goldemberg in his scenario. That relates to a
capacity of about 400 GWe worldwide, slightly more than the above considered
saturation type pessimistic after Chernobyl value of 360 GWe. This implies
also the waiver of CO_2 free uses of natural gas (gas$_2$) though. One then
arrives at 11.05 TW years/year of primary energy use, and also otherwise
numbers become comparable to those of the Goldemberg scenario. It was
explained above that this means a reduction of primary energy uses in the
industrial parts of the world, not of 43% as it was the case in the Colombo
Bernardini scenario but of 69% of the IIASA Low Scenario value for 2030. The
use of 3.7 TW years/year of nuclear power instead of 0.75 TW years/year avoids
such a drastic and unreasonable decrease of primary energy uses and this is
the core of the argument for nuclear power extended here.

In the 1989 Jülich CO_2 reduction scenario 3.7 TW years/year are given
to nuclear power. This is less than in the case of the IIASA Low Scenario
where the total is 5.17 TW years/year. Out of these 3.7 TW years/year, a
share of 2.2 TW years/year is for electricity generation. Assuming a thermal
efficiency of 0.4 (FBR and/or HTR) and a load factor of 0.7 this relates to a
capacity of 1.25 TWe. As an orientation, one may further relate the
1.5 TW years/year of high temperature nuclear process heat to an equivalent
electrical power generation capacity of 0.75 TWe, if again a thermal
efficiency of 0.4 but a load factor of 0.8 is assumed ad hoc. Therefore the
total capacity would be 1.25 + 0.75 = 2 TWe and this is practically the
OECD/IAEA high estimate for 2025. The advantage of this observation is the
fact that all the fuel cycle implications like the ratio between breeders and
burners, the capacities of reprocessing, the consumption of natural uranium
can all be taken from the related OECD/IAEA study [18]. Therefore these fuel
cycle considerations are not repeated here.

Instead, a reflection on the total number of nuclear reactors is in
order. A capacity of 2 TWe means 2000 reactors at 1 GWe each. The capacity
of 1988 which was 331 GWe would relate to 331 such units. Thus an increase by
a factor of 6 is under consideration. It is observed that the safety features
of the 2000 reactors here envisaged for 2030 must be improved by at least a
factor of 6, better more, a factor between 10 and 100, one or two orders of
magnitude.

ad e.)

Energy conservation particularly through improvements of efficiencies are
necessary in any event. In the occasion of the Colombo/Bernardini and
Goldemberg Scenarios this has been discussed at some length. Later in this
paper this will be taken up again, when the technological preparedness or
unpreparedness will be considered (see V.).

ad f.)

It may be possible to think of CO_2 recovery – disposal techniques other than the emission to the atmosphere. In fact, the Marchetti proposal as discussed under a.) already points into this direction. Disposal of CO_2 into ocean currents that aim for the deep sea or the sinking of CO_2 ice into the ocean may be possible. One should keep an open mind in that direction which would permit in particular for uses of coal at a larger scale than otherwise. But more work and ingenuity is required before one can really count on it.

B. Depletion of reserves and resources of primary energy

Besides global warming it was the depletion of reserves and resources that turned out to be a constraint and will therefore be considered now.

The question on reserves and resources is a traditional one for energy analysts. In the past it was overridingly the question of adequate and secure supply that was in the forefront of many considerations. The distinction between reserves and resources is important, in most cases only reserves are being debated. Reserves refer to economic and identified occurrences only, while resources include also sub-economical and undiscovered occurrences. McKelvey has conceived his well known diagram (Fig. 5) to illustrate that distinction [19]. An increasing degree of geological assurance points into the direction of identified reserves and an increasing degree of economic feasibility into that of economic reserves. After the oil shock of 1973 there was a world-wide debate on the adequacy of fossil fuel supply. At that time most of that debate referred to reserves and not to resources.

The assessment of the size of reserves and resources is an ongoing process. It is in particular C.D. Masters of the United States Geological Survey who is engaged there over years. During the last, the XIIth World Petroleum Congress in Houston, he again assessed the size of reserves and resources. Table 12 condenses the results of his evaluation [20] for both, oil and gas. In the case of conventional oil the identified reserves amount to 126.5 billion m^3 or 155 TW years. For natural gas the figure for identified reserves is 110.7 trillion m^3 or 82 TW years for oil and 119 trillion m^3 or 140 TW years for gas. The total of these conventional amounts is then 508 TW years.

As a matter of comparison it is helpful to compare such figures with cumulative uses of fossil fuels in the IIASA Low Scenario by the year 2030. These are 264 TW years for oil and 145 TW years for gas [21], totalling in 409 TW years, which is rather close to the 508 TW years above.

Another comparison may be in order. It relates to the CO_2 content of the preindustrial atmosphere at the beginning of the last century. By then the CO_2 concentration was at 260 ppm, and the total was the equivalent of the burning of 540 TW years of carbon (not coal). The carbon equivalent of the above given 508 TW years of conventional oil and gas is 206 TW years. But one should bear in mind that use of oil and gas in the existing energy systems goes along with the uses of coal and therefore another, say, 194 TW years of carbon equivalent must reasonably be added giving then 400 TW years which is the same order of magnitude as the 540 TW years that relate to the CO_2 content of the preindustrial atmosphere.

Table 12 identifies significant amounts of non-conventional resources and it is in particular shale oil that matters there, 2207 m^3 are indicated. This then would suggest a transition from conventional to unconventional oil resources. In the case of unconventional gas the picture is less clear [22], but there is speculation that these resources may be also very large [23]. The case of the Anadarko basin in Oklahoma must be mentioned here [24]. Therefore the idea of such a transition would also apply to gas. Unconventional resources of oil and gas imply higher costs and partly new technologies. Today, the cases of Athabaska in Canada and Orinoco in Venezuela are examples for that. The estimates of the required production costs are constantly evolving. They may be 40 $/bbl and higher, but it is not possible to make a founded statement here.

But whatever the supply situation turns out to be it is the waste disposal that matters, that is the CO_2 problem. The above given indicative figures point to a remarkable coincidence: The coping with the CO_2 problem for which the 1989 Jülich CO_2 reduction scenario is a drastic illustration suggests cumulative uses of gas and oil that are governed not by the supply side but by the waste disposal side. Table 10 which explained the 1989 Jülich CO_2 reduction scenario suggests a reduced use of oil and an enhanced use of gas by introducing the notion of gas_1 and gas_2 thus essentially doubling the required amounts of gas. Fortunately, Table 12 points out that on top of 110.7 trillion m^3 of identified gas reserves come 119.0 trillion m^3 of still undiscovered conventional resources, more than a doubling. Therefore search for these resources is important and should include as much as possible also unconventional ones. The Anadarko exercise and the recent though unsuccessful Swedish attempt to find such gas in the Siljan Ring area appear thereby in a new light.

The coal resources have not been mentioned so far. Reserves and resources are large and it is purely local conditions and costs that matter. In line with the above given argument one should realize however that during the second part of the next century there could be supply difficulties for natural gas. In that case coal and shale oil would have to be used as sources for the substitution of natural gas (SNG). It is in that spirit that in Table 10 coal was given an index 1 hinting that at a later time it may be necessary to go to a type $coal_2$ when substituting for natural gas_1 is needed.

What is left is a word on uranium supply. In the post Chernobyl situation the installation of nuclear reactors has come to some kind of a saturation. The 1988 capacity is worldwide 330 GWe. In 1988 a post Chernobyl pessimistic evaluation was made that was based on the market penetration analysis using logistic curves [25]. It resulted in an expectation of 360 GWe as a saturation. At such a rate there would be no uranium supply problem. With 30 t of natural uranium per GWe year this would require 1.1 Mio t for 100 years.

The ongoing IAEA reviews [26] result in 3.65 million tonnes of high grade, low cost natural uranium, see Table 13. There is an analogy between fossil and nuclear resources: also in the case of nuclear resources one can identify unconventional ones. Table 13 identifies 10-20 million tonnes of such unconventional resources. The IIASA study suggests even higher values [27] whose uses may however be very problematic. The amount of overburden and refining that is required when the uranium concentration becomes as low as 60 ppm may offset its uses [28]. In any event, if the world's nuclear capacity stalls at 360 GWe the uranium supply question is not a problem for a long time. If, however, nuclear power assumes the role that is expected in the 1989 Jülich CO_2 reduction scenario, not 360 GWe but the equivalent of 2 TWe should be in operation by 2030, a factor of six more [29]. This requires a six times higher annual refuelling and the first core inventories. Fortunately, such a capacity relates nicely to the "high" case of the NEA/IAEA evaluation of "Nuclear Energy and its Fuel Cycle" [18]. Such a nuclear capacity does require the Breeder and a fully established fuel cycle including reprocessing if the requirements for the supply of natural uranium shall be kept below 4 million tonnes. Eventually the Breeder must take over in such a case. The breeding gain of the breeder could be used to sustain also non-breeders, that is Light Water (LWR) or High Temperature Reactors (HTR) [30]. In such a case there exists de facto a decoupling of the operation of such a nuclear capacity or even a larger one for practically infinite periods from the supply of resources. It thus parallels the case of solar power.

The above given considerations for both, fossil and nuclear resources, are fundamentally reassuring so far as the supply is concerned. However, they are mostly of an aggregated kind: locally or regionally there can be severe supply problems. Eastern Europe is certainly a case in point. Its reliance on brown coal poses major supply problems and problems for the environment as well. Given the required brevity of this paper it is not possible to elaborate on that in detail. It is a vast problem of its own.

IV. Time horizons

It is useful now to reflect on time horizons. It helps to organize the thinking and provides orientation. Figure 6 exhibits such time horizons. Regional impacts of the nitrogen oxides and sulphur dioxide emissions establish a problem now. The best recognized case is acid rain but the problems are not restricted to that. The emergence of the CO_2 problem has a time horizon of about 50 years, and it was shown just above that this coincides with the supply horizon for conventional oil and gas. The development of environmentally benign energy systems based on electricity and hydrogen as secondary energy carriers together with nuclear and/or solar power as primary energy carriers have a time horizon of 100-140 years [31]. By contrast, the supply of unconventional oil, gas and coal leads to a time horizon of 250 years or more. The decay of anthropogenic additional CO_2 concentrations by ocean dynamical processes that deliver the CO_2 in an appropriate chemical form to the bottom of the ocean is close to 500-1000 years and there is again a coincidence: also the decay of nuclear waste down to levels of natural deposits is in that order. Finally, supply from nuclear resources for the nuclear breeder (or the fusion reactor) have at least an horizon of 15,000 years or more, and the lifetime of our sun is counted in billion of years.

It is against the background of these considerations of the last four chapters that the technological preparedness or unpreparedness of the decades ahead of us must be evaluated.

V. Technological preparedness and unpreparedness for the years 2020-2050

The years 2020-2050 are closer than one usually assumes. This becomes more obvious when one follows the reasoning and research of C. Marchetti about logistic transitions [32]. Marchetti considers the "market share" F of a given domain (market) into which a technology or any dynamic parameter is evolving. In particular he considers the market share of primary energy carriers besides of many others. (1-F) is then the share of the domain (market) not yet conquered. It turns out that the ratio F/(1-F) then follows an exponential evolution in time, the time constant of which is indicative for its inherent dynamics. In contrast to many other dynamical developments the inherent dynamics of ratios F/(1-F) related to energy markets is long range, covering a century or more. Figure 1 gives such market shares for the consecutive substitution of wood by coal, of coal by oil, of oil by gas, and may be of gas by nuclear and/or solar. There are two remarkable features extended in Figures 1. One is the regularity of the evolutions and the other is their long range. World wars or revolutions occur as local disturbances, the trends are resumed, for instance after World War II.

Not enough is known about such long range dynamical features and Marchetti's analyses are relatively unique or isolated, for that matter. But they sharpen the view of differentiating between top down and bottom up considerations. This is particularly important when the technological preparedness or unpreparedness for the years 2020-2050 is in question.

Accordingly, such a distinction is made in the follows paragraphs when particular technologies or technological problems are being considered.

A. Drilling, deep and cheap (top down)

The primary purpose of drilling in the past has been the exploration and exploitation of oil fields. Technically the oil industry is in very good shape and has demonstrated its capability to supply the countries of the world with sufficient quantities of oil within the rules of the market. These rules require a reserve to production ratio in the order of 10 years and lead therefore typically to bottom up considerations and thereby to strictly commercial-industrial developments. These are very successful. The giant offshore drilling and production platforms in the Gulf of Mexico and the North Sea are good examples for that. As long as an horizon of 10 years (or maybe 20 years) is considered, adequate also from other than commercial viewpoints, governments and public hands are probably well advised to let the oil industry follow its own paths, and that includes the related technological developments. This is probably also true when the transition to non-conventional resources as explained in Table 12 is at stake. Should an energy policy evolve during the next decades that is addressing the CO_2 problem along the lines of the 1989 Jülich CO_2 reduction scenario for the year 2030 as explained in Table 10, such an observation would be even more valid.

The case of drilling for gas may be different. Table 10 considers a doubling of the production of natural gas (gas_1 and gas_2). Even if such a scenario does not materialize, a further increase of gas exploitation is to be expected. Much of it must come from new explorations, and if gas exploration and exploitation are enhanced for reasons like the CO_2 problem that are beyond traditional market mechanisms, drilling for such additional gas becomes important also from a more general point of view. This is top down. Then it becomes apparent that new drilling must be deep and cheap. There is much talk about unconventional gas, the case of the Anadarko well [24] should be mentioned here. Drilling for gas production has been established to depths of 8 km. But also the drilling activities in the Siljan Ring area in the Province of Dalarna, Sweden [33], suggested by the ideas of Gold [23], are a case in point. One should realize that present drilling techniques use a rotating pipe that is operated from the surface having a diameter of typically 20 cm. The twisting then extends over 5–10 km in depth. The transmission of drilling power is in the order of 50 kW and the rotations are slow, 50–100 rpm. The energy that reaches the drilling head is about 10–20%, therefore 5–10 kW. On top of it comes a very limited transmission capability of monitoring data to the surface. Given the perhaps overriding importance of gas drilling also for smaller gas fields from a global overall standpoint, a top down consideration, radically new approaches to drillings, deep and cheap, should be looked for: Downhole motors of a capacity in the order of 500 kW or more capable to operate at elevated temperatures of several hundred degrees, perhaps with fast heads, that is 500–1000 rpm. And above all a much improved monitoring and remote controlled operations that permit also for horizontal detours that avoid obstacles should be pursued. Given the advances in outer space technology that is not an unrealistic goal. Therefore a preparedness for such a development should be looked for.

B. **Long distance transportation of gas (top down)**

Transportation of oil is cheap, in the order of 1–2 $/bbl and therefore not a problem. That is not the case for gas. In the past gas has been transported in pipelines. The cost of such transportation is roughly proportional to the ratio of the length to the diameter of that pipeline. This has led to a limitation of such gas transport in the order of 3,000 km. The power thereby transmitted is in the order of 25 GWyears/year, significant but not truly large. C. Marchetti has evaluated the related international gas trade whose evolution is indeed logistic, that is the ratio F/(1-F) follows an exponential in time. The charm of the logistic analysis is that it permits for an early indication of a later saturation level. Figure 7 thus indicates a saturation level of only 280 x 10^9 m^3/year or 0.33 TWyears/year. Given the 2 + 2 = 4 TWyears/year of the 1989 Jülich CO_2 reduction scenario for 2030 as a yardstick this is insufficient. Figure 7 gives also the time of such saturation: 18 years after the 50% inflicting point of 1976, that is 1994. This is too early, and 0.33 TWyears/year is too little. It is questionable whether larger diameters of long distance pipelines are feasible. Presently 1400 mm is an upper limit for transcontinental pipelines, largely in view of transporting the pipes from the production to the field. Assembling in situ from easily transportable elements may therefore be a future development. Also higher pressures than today's 67.5 – 75 – 80 bars may help but again are probably too difficult to obtain at reasonable costs.

Liquid Natural Gas (LNG) can be transported over larger distances. The point is the lqiuid state of the gas. The best known case is the LNG transport from Indonesia to Japan. Again it was Marchetti who has analysed such LNG transport in logistic terms, (Fig. 8). He arrives at a saturation level of 7.1 million tons/year or roughly 0.008 TWyears/year in 1988. Again this is too little and too early. The cognitive value of such logistic analyses should not be overrated: They can hardly be considered deterministic predictions. Instead, they should be considered as attention marks, bench marks for orientation. The LNG route is expensive and poses safety problems. With the necessary prudence and care these are manageable, the smooth Indonesia-Japan operation is a case in point. But it is unclear whether the LNG route permits for worldwide robust operations in the TWyears/year scale. To that extent a preparedness in that direction should be looked for.

C.   The direct conversion of methane to methanol (top down)

A more elegant route would be the direct conversion from methane to methanol. Direct refers to avoiding the conventional synthesis gas route. This is the route whereby methane is first decomposed to CO and H_2 and these gases are then synthesized to CH_3OH, that is methanol, or higher alcohols, also known as oxygenated fuels. The direct conversion of methane to methanol provides the partial oxydation of the methane in one necessarily well-controlled process step. Its physics and chemistry require research, development and breakthroughs [34]. Probably an appropriate catalyst could do it, also gasdynamical processes have been suggested [35].

Given the importance of natural gas in the time period considered here such liquefaction must be considered as a key element. It would permit for transportation and thus the utilization of natural gas in the TWyears/year scale. It is therefore given a top priority and the prevailing unpreparedness for that technology must be stated here.

D.   Energy storage (top down)

Perhaps the most missing element of future energy systems is cheap and easy energy storage. As a matter of fact, the prevalence of the hydrocarbons may be considered to come from their very good capabilities for rich storing and easy transportation, not from the sheer size of their resources. Their energy content is 10,000 kcal/kg (or 40 MJoule/kg) in the case of oil, a liquid. One must compare that with the energy content of present batteries, 35 kcal/kg or 140 k Joule/Kg [36]. Even permitting for conversion losses during the energy uses of oil it is still a factor of 100 that is in between. This lets the carbon atom appear in a new light: not a source but a mediator of energy, at the cost of relating it to the CO_2 problem!

Energy storage has been attempted in the forms of electric batteries, fly wheels, pressurized air and superconductivity devices [37]. At present their energy density is too small and the costs are too high, tables 14, a - e. This poses a fundamental problem and thus of unpreparedness.

Such unpreparedness is particularly painful in the case of the alternative - or better additive - energy sources like local solar, hydro or portions of wind energy. Again and again it has been stated that their local contributions might be limited but when taken together they could provide a significant share of the overall energy budget. Simply in terms of supply and demand this requires detailed analyses for particular cases. But in any event one must realize the availability problems of such energy sources: There is a fundamental dichotomy in time and space between the availability of such energy and its demand. Bits and pieces of sunshine or wind or small-scale hydro sources must be collected, transported and stored, then they could make more than a marginal contribution. This leads to significant logistic problems: Renewable energies tend to have a power density of less than 1 W/m^2. In the case of biomass a good figure is 0.15 - 0.3 W/m^2. When taking into account the areas of which the rainfall feeds the rivers that is also true for local hydropower. By contrast, the consumption density in cities within their physical, not political limits is always surprisingly close to 5 W/m^2. A city like Vienna covers an area of roughly 500 km^2. For its supply, say on the basis of biomass, that is wood, an area of 10,000 km^2, or 100 km x 100 km, is thus necessary thereby posing significant logistic problems. Indeed, during the times of Maria Theresia, such logistic problems were significant. The energy content of wood was half sacrificed in order to have a light weighted charcoal just in order to overcome the transport problem. All these problems have been overcome by the introduction of the carbon atom as a mediator. This boils down to the problem of easy, light and cheap energy storage.

The substitution of the carbon atom in its role as a mediator for easy collection, transport and storage is an unsolved problem for which we are not sufficiently prepared.

E.   Hydrogen (top down)

The inherently best candidate to substitute the carbon atom in its mediator function is hydrogen. Carbon as an energy source cannot be substituted by hydrogen as it is only an energy carrier. Its production requires energy coming from whatever source. Hydrogen is generally environmentally benign as its residue after burning is water. Its eventual use could be an essential element of a sustainable energy system even for a size of several dozens of TWyears/year. There is hardly any doubt that eventually, that is in the twenty-second century, such a sustainable energy system is to come. It uses hydrogen and electricity as secondary energy and nuclear and/or solar power as primary energy [31], [38].

Given these appealing aspects it is understandable that there exists a lasting enthusiasm for a coming hydrogen age [39]. Yet, there are a number of obstacles that must be overcome first:

(i)   The generation costs

Practically throughout hydrogen comes from electrolysis. Modern electrolysis permits for rather low capital costs and high efficiencies, in the order of 90% [40]. But the electricity costs are still too high [41]. This is inherently due to the fact that electricity itself is a secondary form of energy. If thereby hydrogen becomes a tertiary form of energy it becomes too expensive. This is a fundamental deficit.

(ii) The low density

Hydrogen is of low weight. Therefore it carried 29,000 kcal/kg, a factor of almost three more than oil. But hydrogen is of even lower density, it has 2,560 kcal/m^3 as compared with oil with 8,600 x 10^3 kcal/m^3. This poses a fundamental difficulty. It puts our planes to the forefront of candidates for the future use of hydrogen. That is less so in the case of automobiles. There the storage of hydrogen in the form of hydrids has been investigated. But still it is bulky, heavy and expensive.

(iii) The danger of explosions

There exists the Hindenburg syndrom, it is possible to have hydrogen explosions. Therefore an appropriately engineered safety is an absolute must. One should bear in mind however that an addition of gases other than hydrogen can help. The town gas of the past consisted of more than half of hydrogen and was worldwide in use.

In general it can be stated that a whole hydrogen infrastructure must be created with all the genuine systems problems involved. This poses a long range problem, and what must be looked for are smooth and economically viable transitions from today's existing energy systems to such future sustainable energy systems. Such a transition has been the topic of a whole voluminous book of the KFA Jülich [42]. There is some preparedness for it, and to an extent an unpreparedness. Given the size of the problem it is difficult to cut it into handable pieces. Its appropriate pursuit poses also institutional problems as the time horizon and the globality of the problem are so extended.

F. Solar power, photovoltaic

There is wide agreement that the use of solar power should be pursued. There are now a number of good overviews [43] and these should not be repeated here. Local uses of solar power can at best add to the energy supply. But a completely different thing is the use of solar power at a scale of TWyears/year as for instance pointed to in the 1989 Jülich CO_2 reduction scenario for 2030 where 1.2 TWyears/year are expected, twice the present scale of nuclear energy generation. Such dimension leads to the large-scale, hard technology uses of solar power, mostly in sunny and arid areas of the world, for instance in the mediterranean/North Africa regions. Such options have been studied in greater detail [44]. There is not much doubt that the principal candidate technology for such large-scale uses of solar power is a combination of photovoltaic cells and hydrogen, thereby having in mind that hydrogen overcomes the inherent storage problem as explained above.

Tables 15 a and b provide an overview of the present state of the art.

It has been often expressed as a goal to reach capital costs of 0.5 $US/peak Watt. While one is presently still away from that goal it may be possible to arrive there. Then it is necessary to suppress the systems costs: power conditioning, energy conversion to hydrogen and maintenance. These systems problems will become apparent only gradually as prototype facilities become available and go into operation. What matters are not sizes of 100 kW or 1 MW but 500 and 1000 MWe under reasonable conditions. There are quite a number of research and development activities going on world-wide that relate to the elements of solar power and in particular of photovoltaic cells. To that extent a preparedness for the futures uses of solar power begins to exist. But what is required is more, it is the development and operation of prototype and later demonstration facilities in the hundred MW scale. Only after such a step can the potential of such solar power finally be assessed. Given the size of the costs involved and the lack of public readiness to go into large-scale civilian energy projects there is going to be an unpreparedness.

G. <u>Nuclear energy</u>

At present there is a trend of nuclear energy to saturate at 360 GWe of installed capacity worldwide [25], the present capacity (1988) is close to 330 GWe. At such a level it is not necessary to install the full nuclear fuel cycle. Instead it may suffice to operate in the once through mode where the spent fuel goes to direct disposal. The uranium resources do not create a problem during the next one hundred years or more. It has been already explained above that a significant use of nuclear energy, say at a scale of 2 TWe of installed capacity, is a rather different thing. The use of breeder reactors and accordingly chemical reprocessing is then a necessity.

It must be stated here that practically all countries with the exception perhaps of France are unprepared for the build-up, operation and maintenance of a truly large-scale civilian nuclear fuel cycle. A capacity of 2 TWe requires the operation of 60 reprocessing plants of a capacity of 1,000 t/year. While this is perhaps still conceivable one must imagine the required transportations including the handling of scrap. Lately it was literally the content of 1 mg of Pu that started tremendous regulatory and safety measures in the Federal Republic of Germany. One must realize that at 2 TWe roughly 200 t of fissionable Plutonium are generated each year. Present circumstances suggest that this is beyond the institutional and less so also technical capabilities of single nations. Therefore a number of national fuel cycles adding up to the above-mentioned totals don't appear feasible. Nor do they appear desirable in view of the lasting non-proliferation problems. It may be repeated: The single elements of such a nuclear energy infrastructure may appear to be in trend. But the system of such elements is more than the sum and is essentially unprepared for. What one should try to attempt is the internationalization of the nuclear fuel cycle including the handling and safeguarding of the large amounts of Pu involved and the related waste disposals. Further, what is still to be demonstrated are the large-scale uses of nuclear power for other than electricity purposes, in particular for high temperature process heat, for instance for purposes of methane reformation along the lines of the Marchetti proposal. Systems features and problems that go along with such large-scale uses do have a resemblance to solar power as explained above. So a more new kind of preparedness is therefore being asked for.

H. <u>Energy conservation and improved efficiences (bottom up)</u>

Above it has been explained that energy conservation has to be engaged in any event. Along with it goes the general need for ongoing improvements of conversion efficiences. These topics lead to a long list of a large number of technologies whose improvements are possible and necessary. Practically always this leads to the problem of costs and prices. At an oil price of 18 $/bbl much incentive for such measures has been lost given the mechanisms of market forces. The conservation and efficiency improvement measures are simply too costly. It is therefore more a problem of price incentives, competitivity and regulations that set things in motion. Accordingly only a list of the most prominent elements that are subject to such an improvement shall be given:

(i) The isolation of buildings,
   - the use of heat pumps
   - sophisticated control systems
   - lighting
   - insulations, envelopes

(ii) Transportation improvements
      - improved autombile engines
      - automated traffic control
      - improved aircraft efficiencies

(iii) Electricity applications
      - applications of superconductivity
      - enforced uses of power electronics

(iv) Conversion
      - steam-injected gas turbines and other modern designs
      - use of other power cycles than the internal combustion
      - fuel cells
      - gas cleaning and separation

(v)   Industrial energy efficiencies
      - advanced uses of catalysts
      - advanced heat management
      - co-generation

(vi) Advanced materials
      - ceramics for high temperature engines
      - lightweight structural materials
      - high temperature materials

(vii) Combustions
      - improved computer modelling
      - enhanced fuel switching capability

These are but a few examples and the list is not meant to be complete. But the contrast to the previous paragraphs illustrates the difference between the top down and the bottom up views. It is suggested here that there is a preparedness for the many possible and necessary improvements. They do come by as the market and the necessity develops. As a matter of fact, it is encouraging to look into the past. Over the past 100 and 200 years there has been a regular logistic improvement of efficiencies. Again it was Marchetti who has pointed to that. Figure 9 illustrates that. There does not seem much room for top down interferences.

VI. Recommendations

There are four groups of recommendations reaching from operable technical and economical measures to long-term research and development as well as systems studies.

A. More near-term technical and economical measures

(a) The further operation of coal fired power stations requires the installation of abatement measures for SO_2 and NO_x. This is a particularly pressing necessity in Eastern Europe. The technology for such abatement measures exists. Their implementation is a matter of capital availability and economical agreements. A particular point is the near-term role of brown coal in Eastern Europe. This can no longer be regarded as an isolated problem of the countries in question. It becomes a pressing environmental problem of continental dimensions, a problem of the "Common European House".

It is a problem of today and of high-level political agreements.

(b) Measures for the conservation of energy must be implemented. Again this is less a problem of special technological developments but a problem of price induced market forces, and that is, of economy. Therefore the EC, ECE and CMEA should be instrumental in fostering such inducements. The time horizon for that is the year 2000.

(c) The enhancement of the uses of natural gas includes modern technologies for gas transportation and possibly the reformation to hydrogen as explained above. These are technological problems and the respective industries should be involved to the largest possible extent. Joint ventures of the big oil companies like Shell with partners in the USSR should be looked for. The time horizon for such measures is 2010.

(d) The safety of operating and newly built installations must be looked for. The principal tool for that are probabilistic risk analyses (PRA) as well as managerial and institutional arrangements. The IAEA lends itself there as a principal partner and this requires the explicit will of her member nations.

The time horizon reaches from now to 2030.

B. Necessary, medium- and long-range developments

(a) New technologies for cheap and deep drilling, mostly for natural gas, must be developed. If necessary, this must be done with public support beyond the mechanisms of markets.

The OECD, EC and CMEA should seek appropriate agreements with the big oil companies.

The time horizon for such developments is 2010-2020.

(b) Photovoltaic uses of solar power should not only be researched for but also be demonstrated at a sufficiently large basis which permits the identification and solution of systems problems. This can be done on a national or multinational basis.

The Moscow International Energy Club could be instrumental in the analysis and preparation of such demonstration.

The time horizon for such demonstration is 2000-2010.

(c) Economically meaningful and ecologically viable large-scale uses of biomass should be looked for. This can be done on a national basis with an international exchange of experiences.

The time horizons for that is not well defined, it is an ongoing problem but by 2010 one should know what one is up to.

(d) The nuclear fuel cycle at a large-scale should be internationalized. Only to an extent is this a technological problem. It requires a fresh look on old problems. The Moscow International Energy Club or other such groups could be instrumental there. Only after such preparations can more definite institutional steps be taken.

The time horizon for the internationalization of the nuclear fuel cycle is 2000-2020.

C. Topics for Research and Development

(a) The direct conversion of methane to methanol is still a scientific problem. Given its fundamental strategic importance related R & D should be enhanced. It does not require particular institutional steps but awareness.

(b) New schemes for high density but cheap energy storage without the carbon atom is also still a scientific problem, and related R & D should be enhanced. It does not require particular institutional steps but awareness.

(c) Energy systems using hydrogen should be further researched for and when appropriate, be demonstrated. Awareness for these problems is beginning to exist. Such developments must be pursued with a strategic perspective. Multinational agreements could be helpful.

(d) CO_2 recovery - disposal techniques should be imaginatively looked for in the spirit of geo-engineering.

D. Systems studies

The present energy problems are not a set of isolated well-defined single technical problems. What is at stake is not an improvement of components of the existing energy system but the transition of that system into a new one [42]. This requires ongoing systems analyses. National groups are doing it and it is necessary that they continue, in general more support is required. But in addition a complementary international effort is most appropriate. The International Institute for Applied Systems Analysis (IIASA) at Laxenburg/Vienna among others lends itself for the study of the international and global aspects of the transition with new energy systems.

References

[1] Häfele, W. (Program Leader)
Energy in a Finite World (EIFW)
Vol. 1: Paths to a Sustainable Future
Vol. 2: A Global Systems Analysis
Report by the Energy Systems Program Group of the International
Institute for Applied Systems Analysis
Ballinger Publishing Company, Cambridge, Massachusetts, 1981
See in particular Vol. 2, Chapters 13-20.

[2] See for instance
Leontief, W. et al.
The Future of the World Economy
Oxford University Press, New York, 1977.

[3] See for instance
Hicks, N.L. et al.
A Model of Trade and Growth for the Developing World
European Economic Review, 7, 239-255 (1976).

[4] See Energy in a Finite World (EIFW) [1], Vol. 2, Chapters 14-15.

[5] Frisch, J.R.
- Summary "Energy Balance 2000-2020" of the 12th Congress of the
  World Energy Conference, New Delhi, 1983
  The World Energy Conference, London, 1983
- Future Stresses for Energy Resources
  Graham & Trotman Ltd., London, 1987.

[6] Keepin, B., Kats, G.
Greenhouse Warming - Comparative Analysis of Nuclear and Efficiency
Abatement Strategies
Energy Policy, 16, pp. 538-561 (Dec. 1988).

[7] Colombo, U., Bernadini, O.
A Low Growth 2030 Scenario and the Perspectives for Western Europe
Report prepared for the Commission of the European Communities,
Panel on Low Energy Growth
Brussels, July 1979.

[8] Goldemberg, J., Johansson, T.B., Reddy, A.K.N., Williams, R.H.
An End-Use Oriented Global Energy Strategy
Annual Review of Energy, 10, pp. 613-688, (1985).

[9] Goldemberg, J., Johansson, T.B., Reddy, A.K.N., Williams, R.H.
Energy for a Sustainable World
Wiley Eastern Limited, New Delhi, Bangalore, Bombay, Calcutta, Madras,
Hyderabad, 1988.

[10] Energy 2000 - A Global Strategy for Sustainable Development,
      A Report for the World Commission on Environment and Development
      Led Books Ltd., London and New Jersey, 1987
      See also
      The World Commission on Environment and Development
      Our Common Future
      Oxford University Press, Oxford, New York, 1987.

[11] Toronto Conference Committee
      Statement of the Conference: The Changing Atmosphere,
      Implications for Global Security
      World Conference Toronto, Ontario, Canada, 27-30 June 1988.

[12] Firor, J.
      Public Policy and the Airborne Fraction
      Guest Editorial
      Climatic Change, 12, 2, pp. 103-105, (1988).

[13] Marchetti, C.
      How to solve the CO_2-Problem without Tears?
      Paper presented at the 7th World Hydrogen Conference
      Moscow, UDSSR, 25-29 September 1988.

[14] Niessen, H.F., Bhattacharya, A.T., Busch, M.
      Hesse, K., Zentis, A.
      Erprobung und Veruschsergebnisse des PNP-Röhrenspaltofens in der
      EVA II - Anlage
      Report Jül - 2231, Kernforschungsanlage Jülich
      (August 1988)

[15] Energy for a Sustainable World, [9], Appendix A.4
      pp. 380-384.

[16] Energy in a Finite World, [1], Vol. 2, Fig. 6-3, p. 203.

[17] Energy in a Finite World, [1], Vol. 2, Chapter 23.

[18] OECD/NEA-IAEA
      Nuclear Energy and its Fuel Cycle, Prospects to 2025
      OECD Nuclear Energy Agency and International Atomic Energy Agency,
      Paris, 1987.

[19] McKelvey, V.E.
      Mineral Resource Estimates and Public Policy
      American Scientist, Vol. 60, (Jan.-Febr.), pp. 31-40, (1972).

[20] Masters, C.D. et al.
      World Resources of Crude Oil, Natural Gas, Natural Bitumen and
      Shale Oil
      The Executive Board of the World Petroleum Congress, (ed.)
      Proceedings of the 12th World Petroleum Congress, Houston, Texas,
      April 1987, Vol. 5: Reserves, Finance and General, pp. 3-27
      Chichester, New York, Brisbane, Toronto, Singapore, 1987

[21] Energy in a Finite World, [1], Vol. 2, Table 25-3, p. 782

[22] See for instance
  - Runge, H.C., Fischer, W.
    Erdöl und Erdgas im Übergang der Energiesysteme
    In: Energiesysteme im Übergang unter den Bedingungen der
        Zukunft, [42], Kapitel 2.
    Im Erscheinen begriffen.
  - Hefner, R.A.
    Natural Gas - The Politically and Environmentally Benign Least-Cost
    Energy for Successful 21st Century Economics, the Energy Path to a
    Better World
    In: Meyer, R.F., (ed.), Shallow Oil and Gas Resources, Proceedings
        of the First International Conference, Oklahoma City, July 1984;
        UNITAR Gulf Publishing Comp., Houston, Texas, 1986.

[23] Gold, Th.
    The Origin of Natural Gas and Petroleum, and the Prognosis for Future
    Supplies, Annual Review of Energy, 10, pp. 53-88.

[24] Rice, D.D., Threlkeld, C.N., Vuletich, A.K.
    Character origin and occurrence of natural gases in the Anadarko basin,
    southwestern Kansas, western Oklahoma and Texas Panhandle, USA
    Chemical Geology, 71, 1-3, pp. 149-157, (1988)

[25] Häfele, W.
    Energy Systems in the 21st Century and the Significant Role of
    Nuclear Energy
    (21st Japan Atomic Industrial Forum, JAIF-Conference), 13-15 April 1988,
    Tokyo. Genshiryoku Shiryo (Atomic Information), No. 210, 7, pp. 1-23,
    (1988)

[26] OECD Nuclear Energy Agency - International Atomic Energy Agency
    Uranium Resources, Production and Demand
    1983, 1986, 1988 - Issues
    Organisation for Economic Co-operation and Development, Paris.

[27] Energy in a Finite World, [1], Vol. 2, Table 4-8, p. 119.

[28] Energy in a Finite World, [1], Vol. 2, The case of "yellow coal",
    p. 119 ff.

[29] Häfele, W.
    Global and Regional Energy Scenarios for the Reduction of CO_2
    Emissions and the Role of Nuclear Power
    IAEA/ANL International Workshop on Safety of Nuclear Installations of
    the Next Generation and Beyond, 28-31 August 1989, Chicago, Illinois.

[30] Energy in a Finite World [1], Vol. 2, Chapter 4.

[31] Häfele, W.
    Energy Systems in Transition under the Conditions of Supply and
    Environment
    VGB Kraftwerkstechnik, English Issue, Vol. 68, No. 11, pp. 969-976,
    (Nov. 1988).

[32] Marchetti, C., Nakicenovic, N.
The Dynamics of Energy Systems and the Logistic Substitution Model
International Institute for Applied Systems Analysis, Laxenburg,
Austria, RR-79-13 (1979).

[33] Ingenjörsvetenskapsakademien - IVA (ed.)
Gas - untapped resources?
IVA - Newsletter, December 1987, pp. 11-13
Royal Swedish, Academy of Engineering Sciences.

[34] Gesser, H.D., Hunter, N.R., Prakash, Ch.B.
The direct conversion of methane to methanol by controlled oxydation
Chemical Reviews, Vol. 85, No. 4, pp. 235-244, August 1985

[35] Wilms, M.
Methode zur kinetischen Kontrolle der Methanoxydation
Dissertation in Vorbereitung
Kernforschungsanlage Jülich, Institut für Reaktorbauelemente, 1989

[36] De Luchi, M., Wang, Q., Sperling, D.
Electric Vehicles: Performance, Life-Cycle Costs, Emissions
and Recharging Requirements
Transportation Research - A Vol. 23A, No. 3, pp. 225-278, 1989.

[37] Forschungsstelle für Energiewirtschaft (EfE)
Gutachten für die Enquêtekommission des
Deutschen Bundestages zum Schutz der Erdatmosphäre
Forschungsstelle für Energiewirtschaft, München 1989.

[38] Häfele, W.
Energiesysteme im Übergang unter den Bedingungen
der Zukunft [42], Kapitel 1.

[39] See for example the International Journal of Hydrogen Energy,
Official Journal of the International Association for Hydrogen Energy
Pergamon Press, Oxford, New York, etc., 1975-1989.

[40] Sandstede, G.
Moderne Elektrolyseverfahren für die Wasserstoff-Technologie
Chemie Ingenieur-Technik, 61, 5, pp. 349-361, (1989).

[41] Barnert, H.
Beiträge der Kernenergietechnik zur Wasserstoffwirtschaft
In: Deutsches Atomforum e.V. (Hrg). Jahrestagung
Inforum, Bonn, 1988.

[42] Energiesysteme im Übergang unter den Bedingungen der Zukunft
Ergebnisse einer Studie der Kernforschungsanlage Jülich,
Studienleitung W. Häfele
Im Erscheinen begriffen.

[43] See for instance:
      - The Solar option
        In: Energy in a Finite World [1], Chapter 5
      - Wagner, H.J.
        Nutzung erneuerbarer Energien
        In: Energiesysteme im Übergang unter den Bedingungen
             der Zukunft [42], Kapitel 10

[44] Klaiss, H., Nitsch. J.
      Teilgutachten Erneuerbare Energien für Baden-Württemberg,
      im Aufrag der Landesregierung von Baden-Württemberg
      Materialband: Import solarer Energie
      Deutsche Forschungsanstalt für Luft- und Raumfahrt (DLR)
      Stuttgart, October 1987

References for the figures and tables

Table 9
Edmonds J., Reilly, J.M.
Global Energy. Assessing the Future.
Oxford University Press, New York, 1985

Table 14 a-e
Burkner, W., Hagedorn, G.
Forschungsstelle für Energiewirtschaft
Gutachten für die Enquêtekommission des
Deutschen Bundestages zum Schutz der Erdatmosphäre, [37]

Lorenzen, H.W., Rehm, W., Gröter, H.-P.
Energiespeicherung mit supraleitenden Spulen
Brennstoff-Wärme-Kraft (BWK), Vol. 40, No. 9, pp. 353-360
September 1988

Baumgärtner, K., Kesten, M.
Sichere Druckwasserstoff-Logistik, Speicherung und Transport
Chem.-Ing.-Techn. 56, Nr. 5, pp. 370-376

Fuchs, M.
Wasserstoff in der Energiewirtschaft als Option für die Zukunft?
Das Solar-Wasserstoff-Projekt in Bayern.
Tagungsbericht 8. Hochschultage Energie, pp. 63-79
RWE, Essen, 1987

VDI-GET
Energiespeicherung zur Leistungssteuerung
VDI-Berichte 652, Tagung Köln 4. - 5 November 1987
VDI-Verlag, Düsseldorf, 1987

Table 15 b
Nitsch, J., Steeb, H.
Solarer Wasserstoff: Potential und gegenwärtige Aktivitäten
Spektrum der Wissenschaft, pp. 30-38, August 1989

Figure 1, Figure 2
Energy in a Finite World [1]

Figure 7, Figure 8
C. Marchetti
The Future of Natural Gas
Technological Forecasting and Social Change 31, 155-171 (1987)

Table 1: Comparison of the IIASA Low Scenario for Seven World Regions with the Actual Development 1970-1987, Consumption of Primary Energy [GWyears/year]

Region	1970 IIASA	1970 BP	1975 IIASA	1975 BP	1980 IIASA	1980 BP	1985 IIASA	1985 BP	1987 IIASA	1987 BP
I (NA)	2 363	2 586	2 654	2 706	2 742	2 935	2 830	2 872	2 894	2 953
II (SU/EE)	1 462	1 611	1 835	2 048	2 067	2 458	2 300	2 757	2 435	2 895
III (WE/JANZ)	1 825	1 995	2 256	2 188	2 473	2 397	2 690	2 424	2 783	2 517
IV (LA)	247	268	338	359	449	481	560	542	615	572
V (Af/SEA)	266	361	328	442	449	639	570	811	637	913
VI (ME/NAf)	59	98	126	134	198	168	270	226	309	235
VII (C/CPA)	285	411	461	582	545	735	630	902	677	993
World	6 507	7 338	8 210	8 462	9 030	9 814	9 850	10 536	10 349	11 084

Source: For IIASA - V.G. Chant. Two global scenarios: The evolution of energy use and the economy,
IIASA - Research Report RR-81-35, 1981 For BP: BP Statistical Review of World Energy, 1988

Note: No adjustments of BP Data to IIASA aggregations are made, the date are given as reported by BP.

Table 2: IIASA World Regions I-III

```
------------------
| Region I (NA)  |
------------------
```

North America:
(highly developed, market economy, big energy resources)

Canada
United States of America

```
---------------------
| Region II (SU/EE) |
---------------------
```

The Soviet Union and Eastern Europe:
(highly developed, centrally planned economy, big energy resources)

Albania	Hungary
Bulgaria	Poland
Czechoslovakia	Romania
German Democratic Republic	USSR

```
------------------------
| Region III (WE/JANZ) |
------------------------
```

Western Europe, Japan, Australia, New Zealand, Israel, South Africa:
(highly developed market economy, low energy resources)

Western Europe:

Austria	Luxembourg
Belgium	Netherlands
Cyprus	Norway
Denmark	Portugal
Finland	Spain
Federal Republic of Germany	Sweden
France	Switzerland
Greece	Turkey
Iceland	United Kingdom
Ireland	Yugoslavia
Italy	

Others:

Australia	New Zealand
Israel	South Africa
Japan	

Table 3: WEC-Regions North 1 (N1) and North 2 (N2)

```
North 1 (N1)
```

North America:

Canada
United States of America

Western Europe:

Austria	Luxembourg
Belgium	Malta
Cyprus	Netherlands
Denmark	Norway
Finland	Portugal
Federal Republic of Germany	Spain
France	Sweden
Greece	Switzerland
Iceland	Turkey
Ireland	United Kingdom
Italy	Yugoslavia

Other Industrialized Countries

Australia	Israel
Japan	South Africa
New Zealand	

```
North 2 (N2)
```

Eastern Europe:

Albania	Hungary
Bulgaria	Poland
Czechoslovakia	Romania
German Democratic Republic	USSR

Table 4: Comparison of Population, GDP, and Primary Energy Data of the WEC (L), (M) and the IIASA Low Scenario

North 1, IIASA I + III	WEC 1973 1/	IIASA 1975 1/	WEC 1985 1/	IIASA 1985 2/	WEC 2000 2/ L	WEC 2000 2/ M	IIASA 2000 2/	WEC 2020 2/ L	WEC 2020 2/ M	IIASA 2030 2/
Population [Mio]	783	797	867	868	952		964	1 028		1 082
GDP [10^12 $US,80] 4/	6.56	6.56	8.68	8.94	11.2	12.6	12.1	14.4	18.7	17.5
Primary Energy 3/ [TWyears/year]	83	4.91	5.00	5.29	5.10	5.57	6.70	5.19	6.11	8.91
North 2, IIASA II										
Population [Mio]	357	366	393	393	430		436	465		480
GDP [10^12 $US,80] 4/	2	1.50	1.95	2.34	2.78	3.26	3.92	4.13	5.78	7.6
Primary Energy 3/ [TWyears/year]	1.67	1.83	2.4	2.3	2.98	3.15	3.31	3.49	4.04	5.00

1/ Based on actual data

2/ Based on scenario data

3/ Commercial

4/ Introducing ad hoc the factor 1.62 between 1980 $ for WEC (1973) and 1975 $ for IIASA (1975).

Table 5: Gross Domestic Product, Primary Energy Consumption, and Energy Intensities 1988

Country	Area [10³km²]	Population [10⁶ cap]	GDP/cap [$ 1987/cap]	PEC/cap [toe/cap]	EIC/cap [MWh/cap]	PEC/GDP [toe/10³ $ 1987]	PEC/GDP [Wyr/$ 1987]
FRG	249	61.20	18 264	4.35	6.88	0.24	0.32
France	544	55.63	15 817	3.53	6.45	0.22	0.30
India	3 288	786.00	270	0.29	0.25	1.07	1.43
Italy	301	57.33	13 224	2.58	3.50	0.20	0.26
Japan	372	122.09	18 876	3.09	5.78	0.16	0.22
PR China	9 597	1 077.00	330	0.66	0.44	2.00	2.66
Sweden	450	8.40	18 810	6.62	17.50	0.35	0.47
UK	244	56.93	11 765	3.60	5.32	0.31	0.41
USA	9 363	243.91	18 338	7.58	11.13	0.41	0.55
USSR	22 402	284.00	8 000	5.08	5.86	0.64	0.85
World	135 830	4 980.00	3 370	1.59	2.05	0.47	0.63

Source: Commissiariat à l'Energie Atomique: Mèmento sur l'énergie, 1989.

Table 6: Comparison of final energy [GW years/year] scenarios in the OECD for 1975, 1986 and 2030

	1975		1986	2030		% Reduction a/
	IIASA	OECD b/	OECD b/	IIASA Low	Colombo c/	
Region I(NA), total	1 871	1 823	1 896	2 656	1 819	-32
Industry	757	590	573	1 327	818	-38
Transportation	541	590	662	688	410	-40
Household-service	573	–	–	641	591	– 8
Other sectors	–	643	661	–	–	–
Region III(WE/JANZ), total	1 589	1 487	1 637	3 143	1 723	-45
Industry	805	631	594	1 588	725	-54
Transportation	313	307	415	716	394	-45
Household-service	417	–	–	839	604	-28
Other sectors	–	549	628	–	–	–

a/ Percentage reduction from the Low scenario for the Colombo scenario, as a percentage of the Low scenario.

b/ OECD statistics of 1987.

c/ Based on the evaluation of the "16TW" case in EIFW [1].

Table 7: Comparison of energy scenarios for 2030 (2020), I

Region	Primary energy [TW years/year]			Population [Mio]				
	2030 IIASA Low	2030 Colombo	2020 Goldemberg	1975 IIASA	1987	2030 IIASA	2030 Colombo	2020 Goldemberg
I (NA)	4.37	2.52	)	237	267	315	315	)
II (SU/EE)	5.00	2.98	) 4.3	363	423	480	480	) 1 240
III (WE/JANZ)	4.54	2.45	)	560	601	767	767	)
IV (LA)	2.31	2.23	)	319	424	797	797	)
V (Af/SEA)	2.66	2.50	) 6.9	1 422	1 975	3 550	3 550	) 5 710
VI (ME/NAf)	1.23	1.27	)	133	160	353	353	)
VII (C/CPA)	2.29	2.05	)	912	1 148	1 714	1 714	)
World	22.39	16.00	11.2	3 946	4 998	7 976	7 976	6 950

Table 8: Comparison of energy scenarios for 2030 (2020), II

Region	Per capita consumption of primary energy [kW/cap]					GDP growth rate [%/year] 1975-2030		
	1975 IIASA	1987 BP	2030 IIASA Low	2030 Colombo	2020 Goldemberg	IIASA Low	Colombo	Goldemberg
I(NA)+II(SU,EE)+III(WE/JANZ)	5.82	6.48	8.90	5.09	3.5	2.09	2.00	n.r. a/
IV(LA)+V(Af,SEA)+VI(ME/NAf)	0.43	0.67	1.32	1.28				
VII(C/CPA)	0.5	0.86	1.30	1.20				
IV(LA)+V(Af,SEA)+VI(ME/NAf)+VII(C/CPA)	0.45	0.73	1.32	1.25	1.26	3.24	3.84	n.r. a/
World	2.08	2.22	2.80	2.0	1.6	2.37	2.50	n.r. a/

a/ n.r.: not reported.

Table 9: Carbon content in fossil fuels

Estimated global averages

	Gt C/TWa
Coal	0.751
Liquids	0.605
Gas	0.432

Source: Edmonds/Reilly (1985).

Table 10: A 1989-CO_2-reduction scenario for the allocation of primary energy [TW years/year] to the seven IIASA world regions for the year 2030

Region	Oil	Gas 1	Gas c/ 2	Coal 1	Nuclear d/ 1	Nuclear c/d/ 2	Solar e/	Hydro f/	Bio g/	Total
I(NA)	0.6 (1.18) a/	0.6 (0.67)	-	0.3 (0.69)	0.6 (0.20)	-	0.1	0.12	0.2	2.52 b/ (2.95)
II(SU)	0.5 (0.82)	0.6 (0.88)	0.3	0.3 (1.01)	0.6 (0.08)	0.225	0.075	0.2	0.18	2.98 b/ (2.89)
III(WE/JANZ)	0.6 (1.17)	0.4 (0.37)	0.2+0.2 h/	0.1 (0.53)	0.45 (0.26)	0.15+0.15 h/	0.05	0.07	0.08	2.45 b/ (2.52)
IV(LA)	0.5 (0.31)	0.2 (0.10)	0.3 h/	(0.03)	0.13	0.225 h/	0.2	0.17	0.505	2.23 b/ (0.57)
V(Af/SEA)	0.7 (0.39)	(0.09)	0.7 h/	0.2 (0.34)	(0.03)	0.525 h/	0.2	0.07	0.105	2.50 b/ (0.91)
VI(ME/NAf)	0.2 (0.17)	0.1 (0.07)	0.3			0.225	0.445			1.27 b/ (0.23)
VII(C/CPA)	0.4 (0.15)	0.1 (0.02)		0.6 (0.79)	0.42		0.13	0.17	0.23	2.05 b/ (0.99)
Total	3.5 (4.19)	2.0 (2.20)	2.0	1.5 (3.39)	2.2 (0.57)	1.5	1.2	0.8 (0.73)	1.3	Σ = 16 b/ (11.08)

a/ Values in brackets for comparison with 1987, from BP statistics.

b/ Based on the Colombo scenario.

c/ Assuming a shift from CH_4/H_2O to $CO_2/4H_2$.

d/ Thermal equivalent.

e/ Wind and photovoltaics, direct (electrical) output.

f/ Direct (electrical) output.

g/ Organic waste and plantations.

h/ It may or may not be imported from other regions.

Table 11: Comparison of Primary Energy (PE) in TW years/year and CO_2 Emission (CO_2E) in Gt C/year in Various Scenarios

	1987 a/		IIASA Low 2030		A-1989-Reduction Scenario 2030		reduced nuclear		Colombo 2030		Goldemberg 2020	
	PE	CO_2E	PE	CO_2E	PE	CO_2E	PE	CO_2E	PE	CO_2E	PE	CO_2E
Oil	4.19	2.53	5.02	3.04	3.5	2.12	3.5	2.12	1.72	1.04	3.21	1.94
Gas$_1$ b/	2.20	0.95	3.47	1.50	2.0	0.86	2.0	0.86	0.99	0.42	3.21	1.39
Gas$_2$ b/	–	–	–	–	2.0	–	–	–	–	–	–	–
Coal$_1$ b/	3.39	2.55	6.45	4.84	1.5	1.13	1.5	1.13	4.95	3.72	1.94	1.46
Nuclear$_1$	0.57	–	5.17	–	2.2	–	0.75	–	1.74	–	0.75	–
Nuclear$_2$	–	–	–	–	1.5	–	–	–	–	–	–	–
Solar	–		)		1.2)		1.2)		–	–	0.09)	
Hydro c/	0.73	–	) 2.28	–	0.8) 3.3	–	0.8) 3.3	–	6.60	–	0.46) 2.13	–
Bio b/ d/	–		)		1.3)		1.3)		–	–	1.58)	
Total	11.08	6.03	22.39	9.38	16.00	4.11	11.05	4.11	16.00	5.18	11.24	4.79

a/ BP Statistics, 1989.

b/ Not counting for any losses of methane to the atmosphere.

c/ Direct (electrical) output.

d/ Organic waste and plantations.

Table 12: World Resources of Oil, Natural Gas, Heavy and Extra Heavy Oil, Natural Bitumen and Shale Oil

	Conventional/Non-conventional				
	Conventional Oil [10^9m^3]	Conventional Natural Gas [10^{12}m^3]	Heavy and Extra Heavy Oil [10^9m^3]	Natural Bitumen [10^9m^3]	Shale Oil [10^9m^3]
Production Rate/Year (1985)	2.8	1.5	0.05	0.017	0.0006
Economic					
Cumulative Production	83.3	33.2	5.5	negl.	negl.
Identified Reserves	126.5) 194.0	110.7) 229.7	9.6	0.4	–
Undiscovered Resources (Modal Value)	67.5)	119.0)	–	–	–
Additional Reserves from Future Improvements of Exploitation	18.6 2/	n.a.	–	–	–
Sub-economic					
Identified Sub-economic Resources	–	–	65.8	68.9	522
Undiscovered Sub-economic Resources	–	–	15.1	–	1 685
Ultimate Resources	296 1/	262.9	96.2	69.3	2 207
Oil Qualities:					
Gravity kg/1	<0.934		0.934–>1.0	>1.0	
Gravity °API	>20		20–<10	<10	
Viscosity CP (15°C)			10^3–10^5	>10^5	

(After C.D. Masters et. al. XIIth World Petroleum Congress, Houston 1987, Reprint 25)

1/ Plus 9.3 x 10^9m^3 identified reserves and 10.0 x 10^9m^3 undiscovered resources of Natural Gas Liquids.

2/ C.D. Masters assumes a global weighted exploitation extent of 34% for conventional oil. From tertiary production methods a global 3% improvement of the original oil content of the deposit may be expected.

Table 13: Uranium Resources (10^6 Tonnes U)

	RAR 1/	EAR 2/	SR 3/	ETPR 4/
WOCA 5/	2.315	1.332	6.600-12.100	
outside WOCA				3.300-8.400
	3.647		9.900-20.500	

Source: URANIUM, Resources, Production and Demand, OECD/IAEA, 1986.

1/ Reasonably Assured Resources, up to 130 $/kg.

2/ Estimated Additional Resources, Category 1 (see OECD/IAEA Source).

3/ Speculative Resources.

4/ Estimated total potential resources.

5/ WOCA: World Outside Communist countries.

Table 14 a: Characteristic Dates of Energy Storage Capacities

Type	Stored energy [kWh]	Energy density (mass related) [Wh/kg]	Energy density (volume related) [Wh/dm³]	Cycle efficiency [%]	Spec. Costs $/kWh	Power rel. inv. 3/ [$/kW]
Mechanical Storage						
Pumped-Hydroelectric 1/	$0.5 - 6 \times 10^6$		0.24 (100 m)	65 – 75		500 – 1 500
Compressed Air 1/	1.2×10^6		4.0	40		210
Flywheel	$1 - 10^3$	3 – 45		80 – 90		600 – 1 000 2/
Electrochemical Storage						
Lead-Acid 1/	8.5×10^3 4/	35	85	65 – 75	465	
Sodium-Sulfur		100	130	90	1 100	
Nickel-Cadmium		45	80	55 – 75		
Supercond. Magnet	$0.03 - 1 \times 10^7$		07 – 11.5	90%	275 2/	
Chemical Storage 5/						
Gasoline		12 100	9 120			
Methanol		5 600	4 420			
Hydrogen (liquid)		33 300	2 360			

Sources: Burkner, W., Hagedorn, G., FfE (1989), VDI-GET 652 (1987).

1/ Existing plants in the Federal Republic of Germany

2/ Target, $US 1989.

3/ Conversion factor $US/DM (1989): 2:1.

4/ BEWAG – Plant Berlin.

5/ Net caloric values.

Table 14 b: <u>Characteristic Dates of Electrochemical Energy Storage</u>

Battery-Type	Energy density (per mass) [Wh/kg]	Energy density (per volume) [Wh/dm^3]	Cycle-Efficiency [%]	Spec. costs 1/ [DM/kWh]
Lead-Acid	35	85	65 - 75	930
Sodium-Sulfur	100	130	90	2 200
Nickel-Cadmium	45	80	55 - 75	
Nickel-Iron	45	85	50 - 70	
Zinc-Bromine	65 - 75	80 - 90	50 - 70	

<u>Source</u>: Burkner, W., Hagedorn, G., FfE (1989).

1/ Units in operation.

Table 14 c: **Flywheel Storage Systems**

No.	Type	Stored energy [kWh]	Energy density (per mass) [Wh/kg]	Power related inv. costs [$/kW] */
		Stationary Operation		
1	Garching (1974)	10^3	4.35	
2	Fa. Aerospatiale (1989)	1	2.85	
3	University Ottawa (1988)	8.5	45.7	
4	Fa. Magnet Motor (1986)	300	36.6	
		Mobile Operation		
5	Fa. Garret 1975	3.2	42.7	
6	Fa. Oerlikon 1950/60	5.6	5.6	
7	Fa. Magnet Motor 1977	2.75	n.a.	
		Proposals		
8	Post 1973	20×10^3	100	110
9	Rockwell 1975	4.1×10^3	56.5	550

Sources: Burkner, W., Hagedorn, G., FfE (1989).
Reiner, G., VDI 652 (1987).

*/ $US 1975, target figure 1989: 600 - 1 000 $US/kW.

Table 14 d: **Superconducting Magnetic Energy Storage (SMES)**
Planning (P), under Construction (C), in Operation (O)

No.	Country	Status	Stored Energy [kWh]	Energy Density [Wh/dm^3]	Spec. Costs */
1	Japan	1979 (O)	0.028	2.76	
2	Japan	1980 (O)	0.056	1.77	
3	United States	1979 (P)	0.111	0.69	
4	Japan	1985 (O)	0.111	3.98	
5	Japan	1983 (O)	0.139	2.76	
6	Japan	1980 (O)	0.833	2.76	
7	United States	1981 (O)	5.560	6.22	
8	United States	1983 (O)	8.33	0.87	
9	Japan	1982 (O)	10.83	11.5	
10	United States	1975 (P)	27.8	1.28	
11	Japan	1986 (C)	27.8	7.07	
12	United States	1978 (O)	135.6		
13	Japan	1985 (C)	1×10^3		30 000 1/
14	Japan	1982 (P)	1×10^4	1.60	600 2/
15	United States	1985 (P)	1×10^7	2.76	58 2/
16	United States	1981 (P)	1.2×10^6	1.94	120 2/
17	Japan	1985 (P)	5×10^6	2.65	220-300 3/
18	United States	1980 (P)	5.5×10^6		73 2/
19	United States	1980 (P)	5.5×10^6	5.42	73 2/
20	United States	1978 (P)	1×10^7	2.23	30.4 4/

Source: Lorenzen, H.W., Rehm, W., Gröter, H.-P. (1987).

*/   Target figure 1989: 275 $/kWh.

1/   Cost estimate 1981.

2/   Interpolated from cost estimates 1975.

3/   Cost estimate 1985.

4/   Cost estimate 1978.

Table 14 e: __Hydrogen Storage__

Type	Energy density 1/ (mass related) [Wh/kg]	Energy density 1/ (volume related) [Wh/l]	Energy related 2/ inv. costs [$/kWh]
High pressure vessel 200bar	470	470	12 - 7.5
Al coated pressure vessel 200bar	800	470	12 - 7.5
Steel compound vessel 160bar	2 730	470	22 - 5
Kryo - H_2 - vessel	1 070	600	11.5 - 3.5
Liquid hydrogen storage	5 530	900	11.5 - 3.5
Fe - Ti - hydrid	400	800	23.5 - 13.5
Mg - hydrid	900	800	23.5 - 13.5

Sources: for 1/ Chem.-Ing.-Tech. 56, Nr. 5, pp. 370-76 (1984).

for 2/ M. Fuchs, 8. Hochschultage Energie, RWE (1987), pp. 63-79; $/DM conversion factor: 2:1.

Table 15 a: State of the Art in Solar-Technology (1989)

Collector type	System costs	Energy costs [DM/kWh]	Conditions, installations
Low temperature heat			
unpaned	100 - 250 DM/m^3	0.03 - 0.98	FRG-System efficiency: 200 - 400 kWh/m^3 x year
1 pane	1 000 - 1 900 DM/m^3		EC: 200 - 300 x 10^3 systems $\simeq$ 2.4 Mio m^2
Thermosiphon	1 300 - 1 800 DM/m^3		$\simeq$ 0.17 Mio tce/year $\simeq$ 280 MW$_{th}$
Highly efficient	2 500 - 3 000 DM/m^3		FRG: 20 - 30 x 10^3 systems $\simeq$ 260 x 10^3 m^2 ; $\simeq$ 0.02 Mio tce/year $\simeq$ 30 MW$_{th}$
High temperature heat			
Solar farm plant	4 000 - 25 000 DM/kW$_{el}$	0.14 - 0.16 1/	World: 12 KW$_{th}$ + 5 MW$_{el}$, California 1988: 200 MW31
Solar tower plant	4 000 - 10 000 DM/kW$_{el}$		World: 16 MW$_{el}$
Photovoltaic cell	7 000 - 30 000 DM/kW$_p$ 2/	2.00 - 3.60 2/	Silicon as basic material, energy efficiencies between 7 - 22% ; World: 150 MW$_p$ $\simeq$ 320 GWh/year ; EC: 4 - 5 MW$_p$ $\simeq$ 5 GWh/year ; FRG: 1 MW$_p$ $\simeq$ 1 GWh/year

Source: Wagner, H.-J., Chapter 10 in [42].

1/ California (1988).

2/ Federal Republic of Germany (1989).

Table 15 b: Cost Estimates for Photoelectricity and Hydrogen from Solar Power for the next Decades

Time scale	Investment costs [DM/kW]			Electricity/hydrogen costs [DM/kWh]		Conditions
	Photo-voltaics	Electric Plant	Electro-lysis	Central Europe	Sunny Regions	
1988	12×10^3	30×10^3	1.3×10^3	2.50/3.65	1.30/1.90	MW – plant of present technology
1995	5×10^3	10×10^3	1.3×10^3	0.85/1.30	0.45/0.70	Present technology; Cost degression by higher production: 35 MW/year
2000	3×10^3	7×10^3	1×10^3	0.60/0.90	0.35/0.50	Limits of present technology Production capacity: > 1 000 MW/year
2010	2×10^3	4×10^3	800	0.35/0.55	0.20/0.25	Technology to be developed: Thin layer cells, progress in electrolysis, bulk-production, Capacity: > 1 000 MW/year
2020	1×10^3	2.5×10^3	700	0.20/0.30	0.12/0.12	Supposed attainable limits: Fully established solar industry Capacity: > 10 000 MW/year

Source: Nitsch, J., Steeb, H., Spektr. D. Wiss., Aug. 1989.

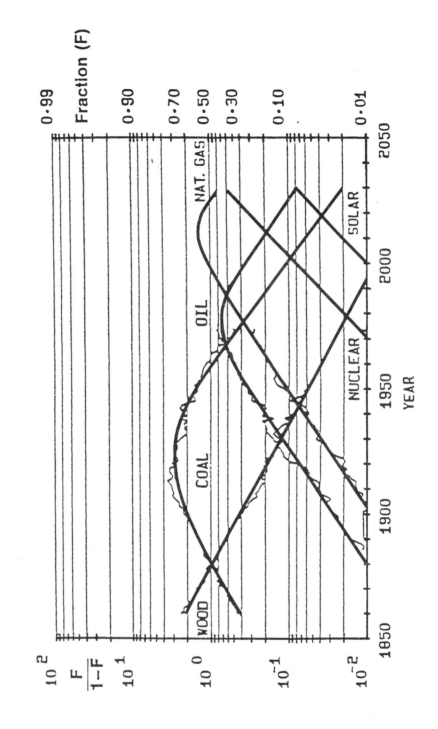

Figure 1: Primary Energy Substitution in the World, 1860 to 2030

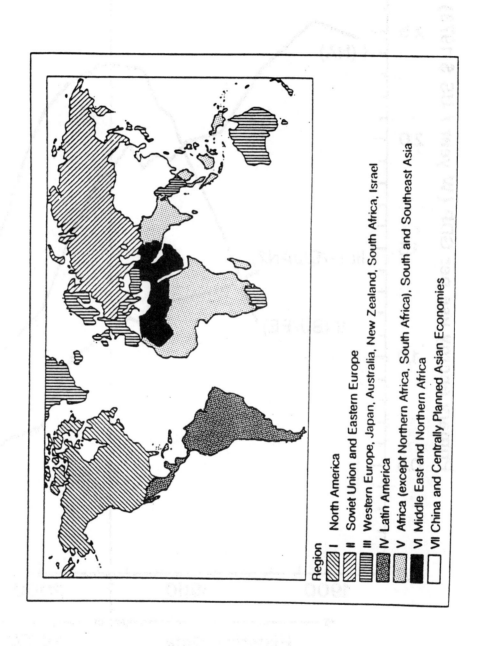

Figure 2: The seven IIASA Regions

Region
I   North America
II   Soviet Union and Eastern Europe
III Western Europe, Japan, Australia, New Zealand, South Africa, Israel
IV   Latin America
V   Africa (except Northern Africa, South Africa), South and Southeast Asia
VI   Middle East and Northern Africa
VII China and Centrally Planned Asian Economies

Figure 3: Energy-to-GDP ratios in the 16 TW case for the IIASA world regions.

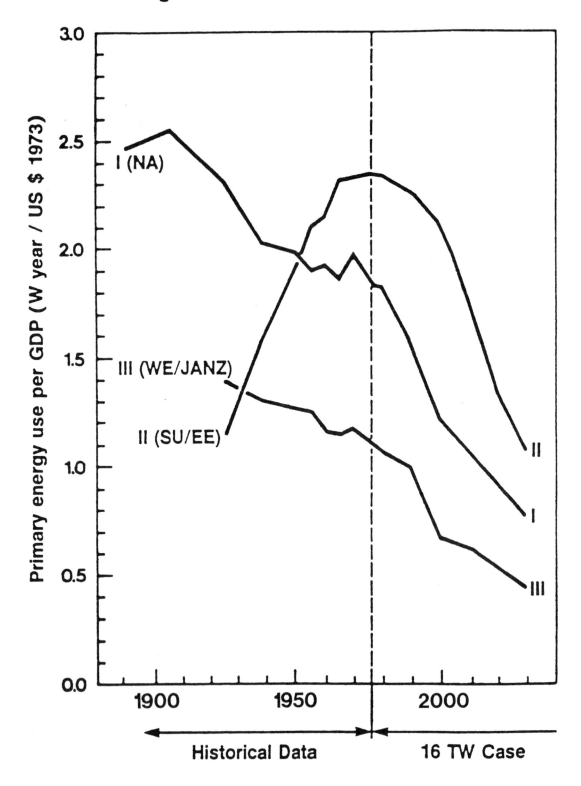

Source: Colombo/Bernardini (GDP in US $ 1973)

Figure 4: Energy-to-GDP ratios in the IIASA Low scenario for the IIASA world.

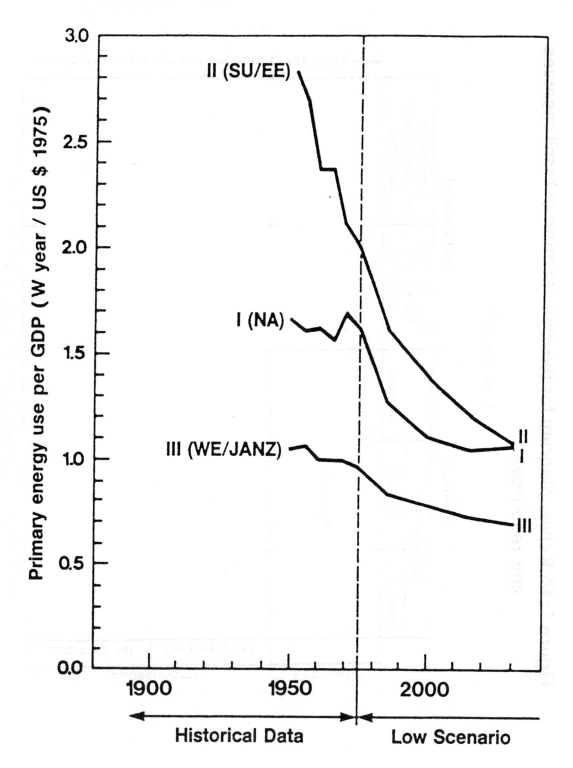

Source: Energy in a Finite World (GDP in US $ 1975)

Figure 5: McKelvey diagram for the classification of fossil reserves and resources

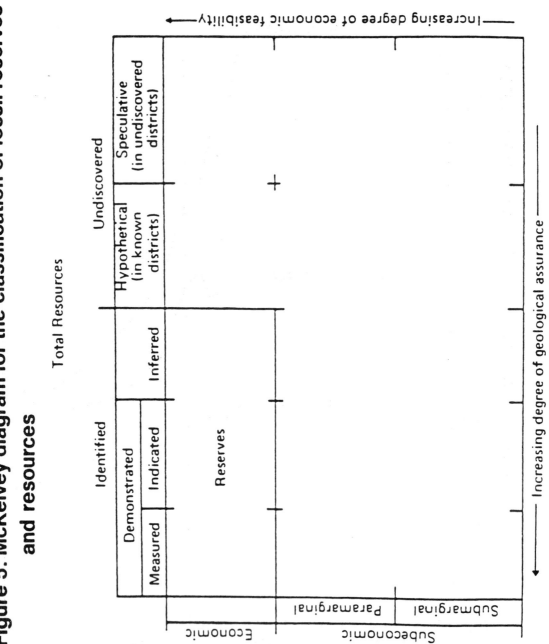

Figure 6:   **TIME HORIZONS */**

Regional impacts of NO_x and SO_2-emissions	now
Emergence of the CO_2-problem (doubling of the CO_2-content)	50 years
Supply from conventional resources (oil + gas)	50 years
Development of environmentally benign energy systems	100 - 140 years
Supply of non-conventional resources (oil + gas) and coal	250 years
Decay of additional CO_2-emissions in the atmosphere	500 - 1,000 years
Decay of nuclear waste down to levels of deposits of natural resources	1,000 years
Supply from nuclear resources (breeder, fusion)	15,000 years
Supply from solar radiation	billions of years

---

*/   Rough estimates only.

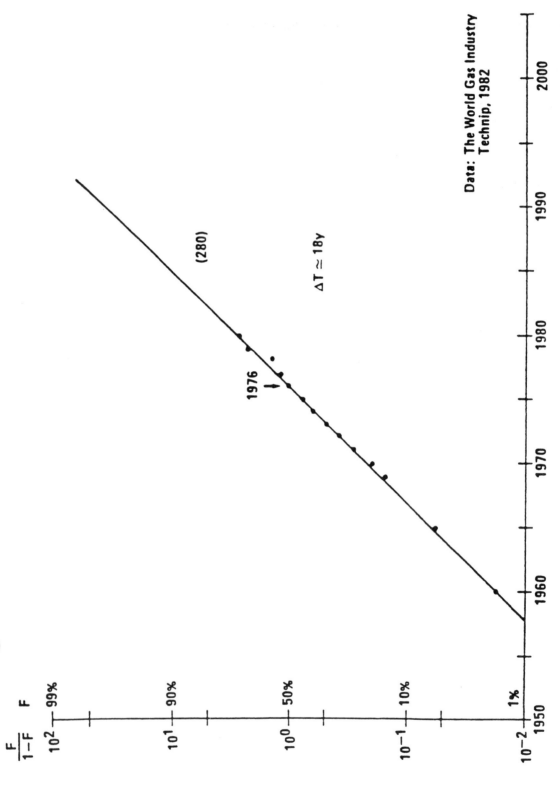

Figure 7: WORLD – INTERNATIONAL GAS TRADE (10^9 m^3)

(280)

ΔT ≃ 18y

1976

Data: The World Gas Industry
Technip, 1982

Source: C. Marchetti, Technological Forecasting and Social Change, 31, 155 – 171 (1987)

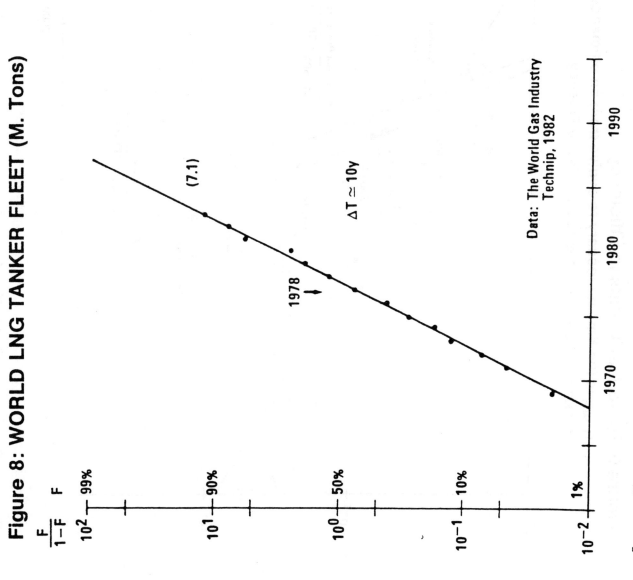

Figure 8: WORLD LNG TANKER FLEET (M. Tons)

$\frac{F}{1-F}$

(7.1)

1978

$\Delta T \simeq 10y$

Data: The World Gas Industry
Technip, 1982

Source: C. Marchetti, Technological Forecasting and Social Change, 31, 155 – 171 (1987)

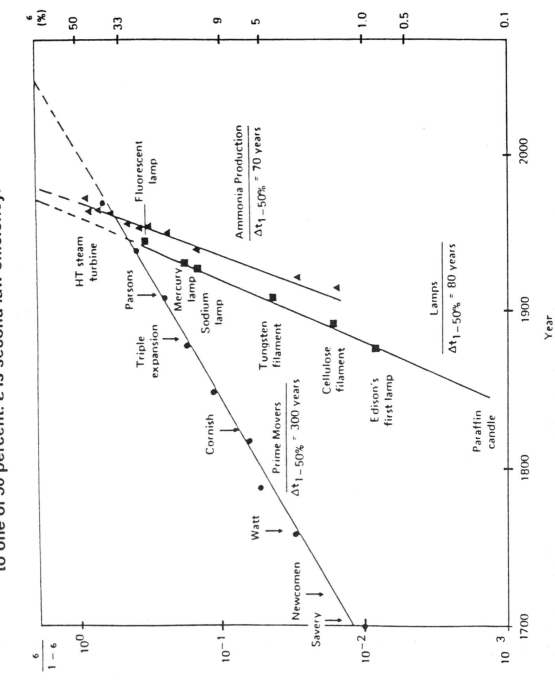

Figure 9: Historical Trends in Efficiency

$t_{1-50\%}$ is the time necessary to evolve from an efficiency of 1 percent to one of 50 percent. ε is second law efficiency.

Source: Energy in a Finite World

Technological Preparedness and Unpreparedness
for the years 2020 - 2050

Energy Conservation and improved efficiencies (bottom up)

(a) The isolation of buildings

(b) Transportation improvements

(c) Electricity applications

(d) Conversion

(e) Industrial energy efficiencies

(f) Advanced materials

(g) Combustions

Topics for consideration

- Drilling, deep and cheap (top down)

- Long distance transportation of gas (top down)

- The direct conversion of methane to methanol (top down)

- Energy storage (top down)

- Hydrogen (top down)

- Solar power, photovoltaic

- Nuclear energy

- Energy conservation and improved efficiencies (bottom up)

SCOPE AND CONDITIONS FOR ENHANCING ENERGY EFFICIENCY
IN THE ECE REGION

1.   The Economic Commission for Europe has had an active energy conservation
programme since 1972.   A major study prepared then entitled "Increased Energy
Economy and Efficiency in the ECE Region" (E/ECE/883/Rev.1) led to Ad hoc
Meetings on energy economy and efficiency held in 1976, 1977, 1978 and 1983.
These Ad hoc Meetings held under the auspices of the Commission were
complemented by major studies, hundreds of technical documents and over
20 symposia, undertaken by the various principal subsidiary bodies of the
Commission.

INTRODUCTION: THE ECE ENERGY CONSERVATION PROGRAMME

A.   ITS DEVELOPMENT

2.   The early ECE activities on energy conservation have covered all fuels
and energy consuming sectors.   They consisted largely of major studies, Ad hoc
Meetings and symposia.   Each seminar or symposium normally produces some
60 technical reports and is attended by about 100 to 150 experts from
Eastern Europe, Western Europe and North America.   The symposia have been held
under the auspices of seven different ECE Committees responsible for a fuel
industry or an energy using sector such as industry or buildings.

3.   The Committee on Housing, Building and Planning held the Seminar on The
Impact of Energy Considerations on the Planning and Development of Human
Settlements in 1977.   The Committee on Electric Power held the Seminar on the
Combined Production of Electric Power and Heat in 1978.   During the same year,
the Timber Committee held the Seminar on the Energy Aspects of the Forest
Industries.   In 1979 the Symposium on the Gasification and Liquefaction of
Coal was held by the Coal Committee and the Senior Advisers on Science and
Technology held the Seminar on Co-operative Technological Forecasting: Solar
Energy.

4.   The efficiency of the production of all fuels was reviewed by the newly
formed Senior Advisers to ECE Governments on Energy in the Symposium on the
Extraction of Primary Forms of Energy held in 1980.   That year the Senior
Advisers on Energy also considered analytical methods for looking at potential
developments in the future in the Seminar on Energy Modelling Studies and
their Conclusions for Energy Conservation and its Impact on the Economy.   The
Steel Committee held the Seminar on the Energy Situation in the Iron and Steel
Industry in 1981 and later held an Ad hoc Meeting on the Strategy for Energy
use in the Iron and Steel Industry 1982.

5.   Energy efficiency throughout the energy system was analysed by the Senior
Advisers on Energy in the Symposium on the Comparative Merits of Energy
Sources in Meeting End-Use Heat Demand in 1982.   The Chemical Committee held a
Seminar on the Rational Use of Crude Oil by the Chemical Industry in 1982.
The Senior Advisers on Energy held a Symposium on the Rational Utilization of
Secondary Forms of Energy in the Economy, particularly in Industry in 1983.

B.   RECENT ACTIVITIES

6.   The study "An Efficient Energy Future: Prospects for Europe and North America" was reviewed by an Ad hoc Meeting on Energy Conservation held by the Senior Advisers on Energy in 1983. Recommendations of the Ad hoc Meeting called for studies on energy efficient technology and a symposium on the Long Term Impact of Energy Efficiency Improvements which was held in 1987. In preparing the symposium the ECE secretariat worked closely with the United Nations Development Programme (UNDP) and the United Nations Industrial Development Organization (UNIDO) on energy conservation in industry, notably in the development of the "Manual of Energy Efficient Technology".

7.   This manual was published by the United Nations as part of a major study entitled "Energy Efficiency in European Industry" (United Nations 1989) and contains some 80 examples of energy efficient technology which is commercially available today from Western and Eastern European countries. Each technology is described by its technical details, state of development, cost effectiveness and specific consumption. Details of the manufacturers and distributors of each technology are given to assist potential buyers or specifiers of this technology to make direct commercial contacts.

8.   While there are thousands of examples of energy efficient technology available throughout Europe and North America today, the ECE manual serves to demonstrate one mechanism for identifying and cataloguing specific examples of energy efficient equipment and disseminating them in ECE member states. In 1989 the Senior Advisers held the symposium on the Optimum Use of Primary Energy Resources in Final Heat Consumption to review efficiency improvements throughout the energy system.

9.   Today there are 36 projects on energy conservation in the work programmes of the Principal Subsidiary Bodies or Committees of the Economic Commission for Europe. But these energy conservation activities are only part of the ECE Energy Programme. One average, every year 20 meetings and 3 to 4 seminars are held at the policy and experts level in the fields of coal, gas, electric power and general energy. Some 15 to 20 studies and surveys are published. Annual statistical bulletins are published in each of the four fields of activity.

C.   ENERGY EFFICIENCY 2000 - A NEW INITIATIVE

10. In future activities considerable attention should be devoted to the fact that east-European energy economies are half as energy-efficient as those of the market economies. To a large extent, this "gap" simply reflects inefficient energy use, but there are other reasons as well. Efficiency gaps also exist among market economies.

11. Reducing this gap by, say half, would save 540 million tons of oil equivalent (Mtoe) in 2000 and 600 Mtoe in 2010, of which 90% would be fossil fuels. Harmful emissions of SO_2 and CO_2 would be reduced by 20-25% in eastern Europe and by about 10% in the ECE region. A reduction of CO_2 emissions of about 10% in the ECE region translates, all other things being equal, into a 5 to 6% reduction of global CO_2 emissions. 18/

12. An initiative, called ENERGY EFFICIENCY 2000, would bring together West and East European businessmen and engineers in some 40 seminars on East-West Energy Efficient Technology Trade and Co-operation per year from 1991 to 1993 and service four symposia on the subject with adequate resources to insure the participation of experts from European IPF 19/ countries.

13. ENERGY EFFICIENCY 2000 would expand the scale of ECE energy conservation activities by an order of magnitude. It would provide a range of services to member states which can contribute to the management of environmental and climatic threats. It is based on extensive experience in this field and on the unique role of ECE in international energy trade and co-operation. It would be executed by the following methods, institutions and financial mechanisms.

(a) Work methods

    1. Hold a large number of workshops in each energy using sector - industry, transport, buildings - under the auspices of national professional associations;

    2. Elaborate a directory and data base accessible in printed and computer on-line form:

        - addresses of individuals in relevant enterprises, banks, government offices;

        - descriptions of technologies, products, services;

        - abstracts of current research;

        - guide to legal instruments, grants, loan schemes and other non-technical measures.

(b) Actors

    1. ECE Commission: elaborates, approves, evaluates project;

    2. ECE Governments: select host institutions; advise on policies, legislation, incentives; contribute to project funding;

    3. Host institutions: provide forum, facilities;

    4. Project secretariat: plans, co-ordinates, services workshops: develops directory;

    5. ECE secretariat: initiates project; provides substantive, administrative support.

(c) Financing

    1. Host institutions: (incremental) costs of workshop facilities partly offset by increased attendance (fees).

2.   Participating ECE Governments:

- secondment of staff/consultants to project secretariat;

- co-financing of project;

- cost partly offset by revenue from sale of directory.

14. The ENERGY EFFICIENCY 2000 initiative would help to achieve the large improvement in energy efficiency between 1985 and the year 2010 shown in the projections prepared for the Study on the Interrelationships between Environmental and Energy Policies.

CHAPTER I: PAST ACHIEVEMENTS, FUTURE POTENTIALITIES

15. Energy efficiency improvement is measured in terms of "energy intensity" or energy consumption in tons of oil equivalent (toe) per $1,000 of Gross Domestic Product (GDP) or Net Material Product (NMP).

16. But while the energy intensity of an economy measures energy efficiency in the broadest terms, it can be influenced by many important features of an economy such as the structure of production, energy prices, government energy conservation policies, the relative share of energy intensive industries, physical activity levels and the actual efficiency of energy production, distribution and end-use.

17. The changes in energy intensity that have taken place between 1973 and 1985 and those projected for the future are given below. Energy use per unit of economic output fell in all ECE countries between 1973 and 1985 except in Poland and in Southern Europe. The increases in Southern Europe were caused by the continued process of industrialization while in Poland an economic crisis took place which no doubt halted further gains in energy

	1973	1985	2000 M	2000 L	2010 M	2010 L
North America	.77	.59	-	-	-	-
Western Europe	.43	.37	-	-	-	-
Market Economies	.59	.47	.38	.39	.33	.34
Eastern Europe	.99	.77	-	-	-	-
Soviet Union	1.07	.98	-	-	-	-
Planned Economies	1.03	.92	.71	.80	.60	.71
Total ECE Region	.67	.57	.46	.48	.40	.43

efficiency over the period. Also, eastern European economies tend to be more energy intensive than in Western Europe because of their greater use of solid fuels, structure of production, less efficient processes, older machinery and slower diffusion of technological progress. 20/

18.   In the future, energy efficiency is projected to improve more and at a greater rate than in the past. Between 1973 and 1985, the energy intensity of ECE countries fell by 15%. During the next 15 years, the energy intensity of the ECE region is projected to fall by about 20% and by the year 2010 it would be 30% lower than in 1985.

19.   In order to achieve this reduction in energy intensity, the large potential for technical and managerial energy conservation measures would have to be harnessed. Some countries have demonstrated how effective a national energy conservation programme can be in levelling off the growth or even reducing sectorial and national energy demand. They see further energy savings as the best preventive method of reducing environmental pollution. While other countries may share this general objective, they are also relying on unprecedented levels of energy efficiency improvements in order to achieve their economic objectives (see Chapter IV below).

20.   This paper shows that the energy savings assumed in this study are technically feasible while potential efficiency improvements could be larger if the best available technology were widely applied. Technical energy efficiency improvements have already had a strong effect on lowering energy demand quite apart from the effect of structural changes, interfuel substitution or changes in activity levels. In the industrial sector, this can be illustrated by changes in total final energy consumption (Petajoules or 10*15 Joules) in the manufacturing industry of the Federal Republic of Germany, Italy, the United Kingdom and France. 21/

	Total Final Energy 1979	Annual Change 1979-84	Annual Change 1983-84	Total Final Energy 1989
Fuels:	6 590	− 375	+ 54	5 120
Efficiency Improvements:		− 253	− 163	
Structural Change:		− 60	+ 79	
Activity Level Changes:		− 54	+ 146	
Interfuel Substitution:		− 8	− 9	
Electricity:	1 554	− 27	+ 64	1 510
Efficiency Improvements:		− 11	+ 2.1	
Structural Change:		− 3.4	+ 18.9	
Activity Level Changes:		− 12.8	+ 42.2	
Interfuel Substitution:		+ 0.1	+ 0.6	

21. The largest reductions in energy demand came from a wide range of energy efficiency improvements. But investments for energy conservation measures linked to investments in new industrial machinery apparently made the largest contribution to these energy savings.

22. Progress in energy efficiency between 1973 and 1985 in the transport sector can be seen in the fuel efficiency of passenger cars measured in litres of petrol per 100 kilometres. 22/

Country	1973	1978	1980	1983	1984	1985
Denmark	9.0	n.a.	8.6	7.3	7.0	7.1
Federal Republic of Germany	10.3	9.6	9.0	8.0	7.7	7.5
Italy	8.4	8.3	8.1	8.0	n.a.	7.8
Netherlands	n.a.	9.2	8.8	n.a.	n.a.	9.1
Sweden	n.a.	9.3	9.0	8.6	8.5	8.5
United Kingdom	11.0	9.1	8.7	7.9	8.8	7.6
United States	16.6	11.8	10.0	9.0	8.9	8.7

23. Passenger transport in North America and Western Europe makes up the majority of transport energy use. In eastern Europe and the USSR freight transport is predominant. Energy efficiency improvements in the transport sector of many western countries has been advanced by government policies, the introduction on more energy efficient vehicles and the relatively quick turnover of the vehicle stock.

24. The effect of energy efficiency measures can also be detected in the building sector by the drop in oil consumption for space heating, measured in Megajoules (Joules 10*6)/dwelling/degree day, in single family homes with central heating in several ECE member States. 23/ Measures include better thermal insulation, double glazing improved central heating appliances interfuel substitution and consumer behaviour. Energy savings in this survey appear to have been largely a result of high oil prices reinforced by government policies to introduce technical measures or "irreversible" conservation measures.

Country	1972	1973	1975	1980	1982	1983
Denmark	52.0	..	42.0	34.5	30.0	30.1
Canada	37.0	38.1	37.4	31.2	27.9	27.5
France	..	52.0	42.5	35.9	..	..
Federal Republic of Germany	47.5	45.5	39.9	30.8	29.4	27.8
Sweden	33.4	..	29.4	22.6	21.7	21.4
United States	71.1	65.4	..	42.2	44.4	..

CHAPTER II. ENERGY EFFICIENT TECHNOLOGIES

25. A wide range of energy efficient technologies are commerically available throughout Europe and North America which could make a large contribution to reducing energy demand. These technologies are manufactured and distributed in both eastern and western countries as described in the ECE Manual of Energy Efficient Technology (ENERGY/SEM.5/3). The potential contribution of different technologies for each energy using sector is described below.

A. Energy extraction, conversion and transportation

26. Over half the losses in the energy system of the ECE region appear to come during the extraction stage. A little over one quarter of losses are at the end-use stage and the rest come mainly from transformation. The present efficiency of the extraction of fossil fuels is relatively low. The average world recovery factor is about 30% of the resouces in place for oil and about 60 to 80% in the case of gas and coal. The related supply potential seems to be especially favourable for oil. 24/ Water flooding of oil fields is of key importance among conventional oil extraction methods. Unconventional extraction methods are expected to contribute only 6 to 10% of production in the United States and the Soviet Union by the year 2000.

27. Relatively little progress has been made on improving the overall efficiency of energy transformation during the last decade. Many countries have implemented combined heat and power systems (CHP) which can reduce primary energy consumption by 25 to 30% in specific cases. But the overall efficiency gains from CHP have been offset in several countries where less efficient coal-fired power plants have taken an increasing share of electricity generation. In the future, combined cycle gas turbines, combined heat and power and district heating schemes can make an important contribution to improving the efficiency of the transformation sector.

28. Combined cycle and steam injected gas turbines for electricity generation are projected to have a 44-48% efficiency by the early 1990s as compared with an average efficiency of 30% today. This would produce overall savings of 32-37% over current average technology.

B. Industry

29. Energy efficient technologies for different end-use sectors representing the current best practice and the best available technology in the foreseeable future are given below. 25/

Energy End-Use Technology	Existing Stock Average Efficiency	New Stock Average Efficiency (% savings)	Best Available Technology Efficiency and Savings	
Iron and Steel	22-24 (GJ/tonne)	17-18 (20-25%)	n.a.	At least 20-25%
Non-ferrous metals (Aluminium)	15-17 (mWh/tonne)	13.5 (10-20%)	n.a.	At least 10-20% (30-40%)
Bricks & Pottery	2.5 (MJ/kg)	1.5-2.0 (20-40%)	n.a.	At least 20-40%
Cement	3.6-3.8 (MJ/kg)	3.3 (8-13%)	n.a.	At least 8-13%
All sectors electric motors	75-95% (% converted to motive power)	80-92% (2-7%)	85-93%	3-12%
Central and on site electricity generation (gas turbines)	30% (% converted to electricity	35%	39-41%	25%

C. Transport 26/

Automobiles	13.0 (1/100 km)	7.9 (39.0)	2.8	78%
Heavy trucks	4.6 (1/100 tkm)	3.04 (34.0)	2.0	57%
Passenger aircraft	53 (1/100 seat km)	28-32 (40-47)	20.0	62%

D. **Buildings**

Energy End-Use Technology	Existing Stock Average Efficiency	New Stock Average Efficiency (% savings)	Best Available Technology Efficiency and Savings	
All electric home	1 242 (Watts/capita)	--	266	78%
Building fabric thermal efficiency	160 (KJ/m^2/dd)	100	50	70%
Space heating oil-gas system	65-70% (% conversion to useful heat)	75-80%	84-94%	23-26%
Refrigerators	1 500 (kWh/year)	1 300 (13%)	750	50%
Water heating	4 000 (kWh/year	3 600	1 700	57%
Service sector space heating-cooling	1.31 (GJ/m^2/year)	0.73 (44%)	0.32	75%

CHAPTER III. APPLICATION OF ENERGY EFFICIENT TECHNOLOGIES

30. Despite the large technical potential for energy efficiency improvements available throughout Europe and North America at present, there are a number of prerequisites to the successful rapid and widespread dissemination of this technology.

A. **Strengthening of market forces and price signals**

31. External factors and barriers in a country may prevent market forces alone from introducing energy efficiency or taking environmental considerations adequately into account. Energy pricing policy, tariffs and tariff structures can be modified to favour fuels with low CO_2 levels, or to reduce energy imports or to stimulate investments in energy conservation. Experience has shown that correct price signals are essential for energy conservation programmes to work well. In addition, there may be any number of barriers to the proper functioning of market forces in energy and environmental matters. These include the inability of manufacturers to fully capture the benefits of research and development work, the risk of developing and disseminating new technology, insufficient access to capital markets, inertia of the market structure and practices of established institutions with counterproductive policies.

B.   Standard setting

32. Providing adequate information to consumers and technical standards to designers and technicians is essential for introducing energy conservation on a broad scale. They help to ensure minimum efficiency levels and to maintain the long-term momentum needed to introduce energy efficiency measures. Standards for automobile emissions, building regulations, labelling of energy using equipment, and energy consumption targets for buildings and industrial machinery provide this information and guide decision-making in the choice and use of energy using technology. Greater harmonization of these national standards among all ECE member States, or at least greater awareness of them would facilitate technical exchanges and accelerate the pace of technology deployment.

C.   Facilitation of technology transfer and trade

33. East-West trade and co-operation in the field of energy conservation makes up a very small proportion of total East-West trade. Joint ventures in heat supply systems, for example, made up only 1% of total East-West joint ventures. In 1987 West-East trade in energy using equipment amounted to about $1,300 million, comparable East-West trade was about $840 million or less than 3% of total East-West trade.

34. Trade and technology transfers would be enhanced by greater comparability of energy efficiency standards, more widely diffused information on energy efficient technology currently available from throughout Europe and North America, and better awareness of exactly who prospective trading partners are in other countries. These are minimum conditions for improved trade and co-operation quite apart from the imperatives for an overall improvement in East-West trade. The conditions for East-West trade, including trade in energy efficient technology, would be enhanced by greater economic decentralization, more responsibility given to individual enterprises and prices which reflect the international market-place (ENERGY/SEM.7/2).

35. This suggests that governments and businesses significantly increase their commitment to East-West co-operation in energy efficiency, in order to help materializing these potentials. Vigorous, innovative approaches are called for.

D.   Availability of capital and multilateral financing

36. Energy conservation has proven to be a highly cost effective investment for central and local government, industries, institutions, building managers, motorists and individual consumers. Coupled with new energy management techniques which cost little, investments in control, measuring, auditing and monitoring technology usually have been repaid in less than two years. Energy conservation investments linked to the replacement of industrial machinery, insulation and efficient heating appliances in new buildings or new motor

vehicle fleets have generally been a highly successful way of introducing energy efficiency improvements, especially when these have been encouraged by government subsidies, grants and loan schemes.

37. Despite the success of these investments, many similar opportunities may be overlooked because of inadequate access to capital markets by potential investors. Third party financing has eased this problem in some countries. Nevertheless, other countries are attempting to execute energy conservation programmes during a period of capital shortage. It would be difficult to extend national or international energy conservation initiatives under these circumstances.

38. Investments for energy management and efficiency improvements have been undertaken in some ECE member States with loans from the World Bank. Other financing mechanisms may need to be considered, including multilateral financing in order to stimulate greater East-West trade in energy efficient technology and implementation of energy conservation measures.

E. Joint research and development

39. East-West exchanges on energy efficient technologies, energy auditing, management and energy policies including research and development take place under the auspices of the Senior Advisers on Energy among ECE member States. Technical exchanges on these matters also takes place through the UNDP-UNIDO Industrial Energy Conservation Network (RER/83/003) which is composed of European IPF countries. 27/ But these exchanges do not include direct joint research and development activities despite the similarity of energy efficient technology applications throughout Europe and North America. Joint research and development activities could reduce the costs of new technology development by eliminating overlapping and duplication of effort in the research community. Such an effort could begin with an East-West energy efficient technology demonstration programme which would document results of recent demonstration projects and circulate them among participating countries.

CHAPTER IV. PRESENT POLICIES

40. Energy conservation is the most common energy policy priority among ECE member States and has been so for at least a decade. At present, energy conservation is considered by many countries to be the best preventative measure for reducing environmental pollution. In an energy policy review carried out in 1988, 17 ECE member States considered energy conservation to be a useful policy measure in treating energy and environmental issues as illustrated below.

Energy Policies of ECE Governments 1988

COUNTRY	PRIORITY POLICIES						ENERGY CONSERVATION												
	ENERGY–ECONOMIC PERFORMANCE	ENERGY PRODUCTION	INTERFUEL SUBSTITUTION	ENERGY RESEARCH–DEVELOPMENT	ENERGY TRADE	ENERGY CONSERVATION	FINANCIAL INCENTIVES	ENERGY PRICING	PUBLIC INFORMATION	INSTITUTIONAL BARRIERS	TECHNICAL SKILLS	DEMAND MANAGEMENT	TECHNICAL SOLUTIONS	INDUSTRY	TRANSPORT	BUILDINGS	TRANSMISSION–DISTRIBUTION	ENVIRONMENTAL ISSUES	REDUCE OIL CONSUMPTION
ALBANIA																			
AUSTRIA																			
BELGIUM																			
BULGARIA																			
CANADA																			
CYPRUS																			
CZECHOSLOVAKIA																			
DENMARK																			
FINLAND																			
FRANCE																			
GERMAN DEMOCRATIC REPUBLIC																			
GERMANY, FEDERAL REPUBLIC OF																			
GREECE																			
HUNGARY																			
ICELAND																			
IRELAND																			
ITALY																			
LUXEMBOURG																			
MALTA																			
NETHERLANDS																			
NORWAY																			
POLAND																			
PORTUGAL																			
ROMANIA																			
SPAIN																			
SWEDEN																			
SWITZERLAND																			
TURKEY																			
UNITED KINGDOM																			
USSR																			
USA																			
YUGOSLAVIA																			

Source: Senior Advisers to ECE Governments on Energy, 1988.
International Energy Agency, 1987 Review.

41. Coincidentally, the energy policies pursued by ECE Governments since the middle of the 1970s generally have contributed to the reduction of environmental pollution. Experience with various policy measures, energy pricing, institutional mechanisms, technology deployment and analytical techniques during the recent past is directly applicable to achieving sustainable energy use. Interfuel substitution or fuel switching away from oil, for example, was and still is an essential element of energy security in many countries in the region. This same approach is now being applied to the substitution to less polluting energy sources. Decoupling economic growth from the growth of energy demand has been an important objective in many ECE countries for over a decade. Some countries now suggest that one of the challenges of the future will be to decouple economic growth from the use of environmentally harmful energy sources.

42. Energy conservation policies have been remarkably successful in many ECE member States. Several countries have noted that while energy consumption was approximately the same in 1987-1988 as in 1973, economic output had increased significantly, in some cases by over 30%. Others noted that energy efficiency had increased by 25% over the same period. Many countries are considering targets for the future within this range, aiming at efficiency improvements of 15 to 20% by the year 2000. Finland, for example, proposes to keep primary energy consumption at the 1989 level to the year 2000 and then lower it by 10% in 2010. On the other hand, a few countries envisage very large energy savings, large both in historical terms and by international standards, improving energy consumption per unit of economic output by 40 to 50% by the turn of the century. Other countries have established targets for pollution levels, such as the reduction of SO_2 emissions from power stations by at least 60% of 1980 levels by the year 2005 and reducing NO_x emissions from power stations by at least 50% of 1980 levels by the end of the century.

43. In order to meet these challenges, most countries are continuing or intensifying their energy conservation programmes and increasing support for research and development especially for energy efficient technologies. In addition, many governments appear to be aware of the need for international co-operation on energy and environmental matters in order that their national efforts have a discernible effect on environmental pollution levels. Many countries engage in international co-operation within the ECE region and with developing countries through a variety of international and intergovernmental organizations, and see the scope for greater co-operation.

44. Energy conservation programmes range from essentially providing information and advice on cost-effective energy conservation measures in some countries to plans in others for economic restructuring away from energy intensive industries or changing the modal split in transport away from cars to increased public transport. The instruments most commonly used are regulations, appliance labelling, loans, grants, energy audits, investment subsidies, accelerated depreciation for tax purposes of energy efficient equipment, market-based incentives, energy pricing and tariffs, consumer education and information, taxes, fines, monetary rewards, and energy technology research, development and demonstration.

45. In the housing sector, several countries are preparing to tighten building regulations and improve co-operation between the energy utilities, architects and city planners. Heat planning will continue to guide the implementation of combined heat and power or district heating and careful

attention will be given in the future to the choice of fuels providing heat to these systems. In _transport_, some countries are encouraging a shift away from road transport towards rail and barges for freight and to improved public transport for passengers. Several countries either have had for some time or are introducing lead-free petrol, catalytic converters and energy efficiency labelling for automobiles. In _industry_, a few countries noted measures to improve energy auditing, energy management and to introduce energy efficient technologies, especially in energy intensive processes.

46. In the _energy production sector_, several countries plan improvements in electric power generation such as the installation of combined cycle gas turbines, fluidized beds, or environmental pollution control techniques to reduce emissions from low calorific value solid fuels. Many countries now require environmental impact assessments for the exploitation of new fossil fuel deposits or hydro potential. Several countries have stopped the expansion of their nuclear power programmes and some of these are preparing to phase out their nuclear power electricity generation capacity. On the other hand, at least one country, France, has noted that CO_2 levels have been reduced significantly, by 38% between 1980 and 1988, as a result of its national nuclear power programme.

47. Research and development programmes for energy efficient and pollution reduction technologies have been strengthened in several countries while others acknowledge the need for greater applied research, development and demonstration, especially in electric appliances. This includes improvements to existing energy efficient technologies such as heat pumps which require a heat pumping medium that does not effect the ozone layer the way Freon, the medium in common use, does now.

CHAPTER V. SUMMARY

48. It follows from the foregoing that international co-operative activities would prove particularly beneficial if oriented towards:

1. Enhancing international trade and co-operation in energy efficient technology by increasing contacts between Western and Eastern European businessmen and engineers, holding seminars, and establishing an international registry of enterprises, manufacturing and commercial companies which are or could become engaged in international trade in energy efficient technology or consulting in energy demand management techniques;

2. Harmonizing national technical standards of energy efficient technology, energy efficiency labelling schemes and, energy performance indicators;

3. Improving the institutional, legal and organizational prerequisites for enhanced east-west trade in energy efficient and environmentally acceptable appliances and equipment;

4. Establishing an international system of data exchange for ECE countries on methodologies and advanced energy efficient technologies based on the format of the ECE Manual of Energy Efficient Technology (ENERGY/SEM.5/3) (see Annex);

5. Providing internationally comparable data on end-use heat demand, by establishing common definitions and classifications.

Questionnaire for the ECE Manual on Energy Efficient Technology

MANUAL OF ENERGY EFFICIENT TECHNOLOGY

1. SECTOR	2. NAME OF TECHNOLOGY, PROCESS OR TECHNIQUE
3. COUNTRY	4. NAME, ADDRESS, TELEPHONE AND TELEX OF SUPPLIER OR MANUFACTURER

5. TECHNICAL SPECIFICATION

6. SPECIFIC ENERGY CONSUMPTION

7. STATE OF DEVELOPMENT OR USE

8. COST EFFECTIVENESS

9. DIAGRAM OR ILLUSTRATION (Please attach)

CO_2 AND CLIMATE VARIATION: ITS IMPACT UPON ENERGY POLICY IN THE USSR AND EUROPEAN CMEA COUNTRIES 28/

INTRODUCTION

The present study was made under a contract with the United Nations Economic Commission for Europe (ECE). It is a part of a broader study concerning the sustainable energy development in the ECE region in the twenty-first century.

The analysis considers opinions voiced by Soviet scientists on the problem of the "greenhouse" effect, on the role of specific "greenhouse" gases and on their effect on global change of temperature and climate.

Along with the assessment of recent development of the Soviet energy industry this analysis evaluates the dynamics and structure of greenhouse gases released to the atmosphere in the USSR. An attempt has been made to estimate also the total amount of these gases in other European countries participating in the Council of the Mutual Economic Assistance.

The major efforts aimed at reducing the pollution of the atmosphere in the USSR, primarily as regards the negative role of power plants are briefly described. They include:

(i) advances in restructuring the energy balance of the country, resulting in a decrease of fossil energy sources with relative high emissions of CO_2 and other greenhouse gases at a burning stage and simultaneous increase of "clean" energy sources;

(ii) development of energy technologies and technical means reducing the amount of greenhouse gases during burning of fossil fuels;

(iii) development of advanced technologies and technical means collecting hazardous components of emission gases;

(iv) increasing the energy efficiency in order to reduce the energy consumption per unit of final product and work spent.

Using forecasts of the industrial development of the USSR towards 2010 (and in some cases specific aspects in the more distant future), an estimation of probable amounts of CO_2 emissions into the atmosphere by energy plants has been made.

On the basis of this study numerous conclusions are drawn identifying global investigations of highest priority as well as areas of industrial, scientific and technical co-operation between the USSR and other countries in this field.

A bibliography on the greenhouse effect and the role of CO_2 is also given.

EXECUTIVE SUMMARY

Main conclusions and recommendations on the development of international co-operation in the area of the greenhouse effect

1.   Scientists and engineers all over the world agree that the progress of the greenhouse effect in the next decades may result in irreversible climatic changes on the earth and negative consequences for human living standards, economy and ecology.   These negative phenomena can only be avoided through mutual international efforts, the USSR, the United States of America, China, West Europe and Japan being the leaders of this movement.   In this situation the development of absolute new economic, political, scientific and engineering co-operation concerning numerous problems related to the greenhouse effect and means to avert its negative climatic, economic and ecological impact is of utmost importance.   This policy should also involve raising the financial and intellectual resources and could be greatly assisted by gradual disarmament.

2.   The development of the greenhouse effect is mainly caused by increasing concentrations of carbon dioxide, methane, nitrogen oxide and several freons in the earth's atmosphere.   The greenhouse effect is presumably already under way; 60% of its action stems from the emission of CO_2, 20% from CH_4 and 15% from CFCs.   Assuming that the nearest future will not be marked by economically acceptable technological solutions resulting in a dramatic decrease of CO_2 emissions, the major international efforts in this area need to be as follows:

   (i)   a substantial reduction in industrial branches having a high energy intensity; an increase of the efficiency of energy utilization and a restructurization of the energy balance in order to decrease (relatively and absolutely) the amount of direct use (without preliminary processing) of fossil fuels (coal, shales) with a high carbon content; an increase in the fraction of low carbon-containing fuels (natural gas); a greater role of atomic power, hydrogen and renewable energy sources (hydroenergy, solar, geothermal, wind, tide and biomass);

   (ii)   restricting production and use of freons;

   (iii)   development, implementation and international trade and co-operation of ecologically clean coal-burning technologies;

   (iv)   development, implementation and international exchange of technologies and technological means reducing to the minimum the leakage of CH_4 at gas and oil-mining sites, during accidents at main and feeder gas pipe lines and underground gas tanks, from coal-mines, urban dumps and rural animal farms.

3.   Problems of climatic changes owing to the greenhouse effect need to be studied in more detail especially with respect to determining the timing and magnitude of these effects and resulting negative impacts on environment, economy and living standard of mankind.   The climatic global system itself, interaction of the ocean with atmosphere, the problem of dynamics of the greenhouse effect after stabilization of concentrations of atmospheric

greenhouse gases, its effect upon various climatic zones and other problems should be also clarified. Along with a rational approach to the above problems studies should be undertaken at the international scale using advanced experimental techniques and equipment according to mutually accepted standardized processing algorithms. International studies must be based upon a special international agreement signed by all (or nearly all) Governments. This work could result in an International Treaty on the Conservation of the Earth's Climate. This treaty could form a basis for large-scale global economic, scientific and engineering procedures to control the growth of the greenhouse effect. This activity should be developed within the framework of the United Nations Organization.

4.   The progress of the greenhouse effect depends to a certain extent on the status of the main forest areas and degree of pollution of the ocean. In this respect a substantial intensification of international economic, scientific and engineering co-operation is necessary. It should involve preservation of the CO_2 absorbing capacity of forests and oceans via the protection and restoration of forests and protective measures in the ocean area primarily aimed at the removal of the oil film from its surface.

5.   Gaseous carbon dioxide is the main contributor in the greenhouse effect. On the other hand it is an important raw material in the production of chlorella - a biological source of nutritional additives for cattle. The USSR and several other countries have already developed the modern technologies to manufacture chlorella. The foundation of an international organization dealing with the development of a large-scale chlorella production as well as the use of CO_2 as a food conservant at energy plants which are the major sources of the above gas appears justified.

6.   The United Nations Organization should adopt a detailed resolution calling upon all countries to intensify studies aiming at:

   (i)   a decrease of the energy intensity per unit of GNP;

   (ii)   the development of energy sources, whose production and consumption does not create greenhouse and other gases and/or dramatically decreases the emission of these gases (hydrogen energetics, non-conventional renewable energy resources, first of all solar energy and its derivatives);

   (iii) ecologically clean fuel technologies, which can be used either at new or at operating plants (solid fuel - steam power plants with emission clean-up; underground and integrated coal gasification; concentration and processing of coal and shales; capture and utilization of methane from dumps; novel technologies of fossil fuel combustion, involving fluidized bed furnaces and new furnace designs; ecologically clean gas processing plants; reduction of evaporation of oil and oil products, of methane emissions from coal-mines, open cuts, damaged gas pipe lines and from accidents at plants of gas industry;

   (iv)   broadening the basis of the gas production by prospecting of new gas fields and utilization of non-conventional deposits (gas hydrates).

7.   <u>A significant role in the reduction of global anthropogenic emissions of</u>
<u>CO_2 can be played by pure economic factors</u>, i.e., increasing the fraction of
natural gas with low carbon content. It is necessary to introduce an index of
"greenhouse safety" in the process of fuel price setting. In other words, the
fuel price should "credit" fuels with relatively low CO_2 emissions.

8.   In order to draw the attention of the population of the European
countries to the problem of an increasing content of greenhouse gases in
the atmosphere and its possible negative impact on the global climate, it
might be appropriate to hold a symposium, possibly under the aegis of the
United Nations ECE in 1990 or 1991 on the interrelation between the
development of the energy economies and energy conservation, and the
greenhouse effect, in the ECE region.

9.   In principle, <u>in the USSR it is possible to slow down the growth rate of</u>
<u>emission of the greenhouse gases by 2000</u> (reduced to the carbon equivalent)
<u>and to reduce their emission into the atmosphere by 25% by the year 2020</u>.
This will necessitate the realization of numerous large-scale projects,
primarily aimed at reducing the energy intensity of the economy and further
restructuring the energy system. The latter would require the development of
nuclear power based on a new generation of safe atomic plants (highest
priority); the increased production and effective utilization of natural gas;
technological upgrading of existing and large-scale building of new steam
power plants with clean coal technology; and mastering of non-conventional
renewable energy resources. The implementation of this extraordinary
economic, scientific and technological task would demand significant material,
financial and labour resources which are not likely to be available in the
present economic situation in the USSR. However, the situation might change
in a few years provided attempts in this direction are supported by all or
nearly all industrial countries in the framework of close co-operation.

CHAPTER I.   GREENHOUSE GASES AND RELATED FACTORS AFFECTING THE EARTH'S CLIMATE

Soviet scientific media define the "greenhouse effect" as heating of the
inner layers of a planet's atmosphere (Earth, Venus, etc., having dense
atmospheres) due to the transparence of the latter to most of the solar
radiation (in the visible range) and to the absorption in the atmosphere of
most of the heat (infrared) radiation emitted by the surface of a planet
heated by the sun.

<u>The Soviet scientific media classifies carbon dioxide (CO_2), methane</u>
<u>(CH_4), nitrous oxide (N_2O), chlorofluorocarbons (CFCs), nitrogen oxides</u>
<u>(NO_x) and tropospheric ozone as greenhouse gases</u> prompting a greenhouse
effect. The greenhouse effect can be also originated by water vapours.

Under normal conditions, i.e., when the amount of greenhouse gases equals
or slightly exceeds the "background" value, the greenhouse effect plays an
extremely important role: it maintains the earth's temperature within the
limits necessary for the development of life and smoothing the difference
between night and day temperatures. According to estimations the average
temperature of the earth's surface is 32° C higher than it would have been
without the greenhouse effect.

The carbon dioxide is being considered as a major component of atmospheric greenhouse gases. Its concentration began to increase in the earth's atmosphere from the end of the eighteenth century due to reduction of the forest area, which naturally led to a decreasing amount of carbon dioxide absorbed by photosynthesis. A further increase in CO_2 concentrations originated from the burning of fossil fuel. The latter is now a major source of growth of concentrations of atmospheric CO_2, which over the past 200 years has increased by about 25% [1]. The main part of this rise was observed recently. The studies revealed that the content of CO_2 in the atmosphere increased by 12-17% in recent dozens of years. Now its annual increase is approximately 0.2%.

The annual consumption of fossil fuel involves $10-12 \times 10^9$ tons of oxygen which once linked to carbon, creates carbon dioxide. Presently, the concentration of CO_2 in the atmosphere amounts to 0.032% (above cities 0.034%). The toxicologists indicate that CO_2 concentrations up to 1% are harmless for human beings and that the suppression of plant life begins only from 2%.

The concentration of methane began rising from the end of the seventeenth century. It has more than doubled over the last 300 years [1].

The major source of atmospheric CH_4 is the decay of fossil substances in the absence of oxygen, the emission of methane with ventilation gases from coal-mines and the leakage of natural gas due to the exploration of mining areas and pipeline accidents. According to several estimations the annual increase in atmospheric CH_4 was about 0.8-1% over the past dozens of years whereas presently it is equal to 1.5%. This gas approximately accounts for one fifth of the total greenhouse effect. Methane is unstable and its lifespan in atmosphere (in contrast to CO_2) is relatively short. However, as regards its greenhouse effect one molecule of CH_4 is equivalent to 20 molecules of CO_2. A slight decrease of the emission of methane into the atmosphere (according to several experts of only 10%) would stabilize its concentration in the atmosphere.

Chlorofluorocarbons are exclusively products of the chemical industry. They cannot generate under natural conditions. Until the early 1950s, there were no chlorofluorocarbons in the atmosphere. Estimations of scientists indicate that the concentration of chlorofluorocarbons is about three orders of magnitude lower than that of CO_2; but the greenhouse effect of each molecule of chlorofluorocarbon entering the atmosphere is 20,000 times as much as that of a single CO_2 molecule. Recent years were characterized by an annual increase of concentrations of atmospheric chlorofluorocarbons of 5-10 per cent [1]. They account for up to 15% of the annual increase of the greenhouse effect. It has been also calculated that the stabilization of concentration of the above gases necessitates a decrease of emissions of $CCl_3F(F-11)$ by 75% and $CCl_2F_2(F-12)$ by 85%. Along with the deterioration of the earth's climate due to the greenhouse effect, chlorofluorocarbons destroy the ozone layer. The rising concentration of chlorofluorocarbons is mainly caused by their extremely slow decomposition (dozens or hundreds of years) under natural conditions.

The major mass of ozone is situated at altitudes of 25-30 km and protects our planet from ultraviolet radiation. The ozone layer absorbs heat radiation. An increase of its temperature results in a warming of the atmosphere due to thermal exchange. Ozone also absorbs heat radiation emitted by the earth. Therefore, this gas, similarly to CO_2 should be classified as a greenhouse gas. It has been suggested that its role with respect to the greenhouse effect is almost the same as that of carbon dioxide [2]. In other words the amount of ozone as well as the concentration of CO_2 affect the earth's climate.

According to experimental data collected by Soviet scientists, the ozone hole over the Antarctic already amounts to nearly $5 \times 10^6 km^2$. The total amount of ozone has gradually decreased in the whole atmosphere, by 3% over the last 10 years, mainly due to the increase of concentrations of chlorofluorocarbons [1]. A minor variation in the content of ozone (1.7%) is observed in the equatorial belt, whereas at the latitude of Moscow the loss amounts to 3%. A certain increase in ozone concentration was observed in the 1970s, but the drop during the winter in the 1980s was 7%.

The ozone depletion is mainly caused by weak stability of its molecules when affected by increasing amounts of chlorine-, bromine- and nitrogen-containing compounds. This can be explained by extensive use of nitrogen fertilizers and halogen hydrocarbons. Chlorofluorocarbons are the most active ozone-depleting compounds. They decompose owing to the effect of solar ultraviolet rays releasing chlorine which destroys ozone. As a result of such processes the atmosphere gets saturated with chlorine, whose concentration at the altitude of 40 km is several times higher than that 20 years ago.

The increase in atmospheric concentration of nitrogen oxides mainly stems from the processes developing in soil to which a wide use of nitrogen fertilizers in agriculture and large-scale combustion of fossil fuel have been added recently. However, the sinks of this gas appear not to be studied at all.

During the past few years the concentration of N_2O has been increasing relatively slowly (about 0.25% per year), whereas the amount of emissions of this gas continuously increased. Numerous scientists note that this discrepancy has not yet been clarified. Nevertheless, it has been established that the greenhouse effect per molecule of N_2O is 10 (and according to other estimations – hundreds) times higher than that of a single CO_2 molecule. Morever, the effect of N_2O (similar to chlorofluorocarbons) has an unfavourable impact upon the stratospheric ozone layer. According to estimations of several experts its concentration has increased by 10% over the past century. In order to stabilize the amount of N_2O at present levels its emission must be reduced by no less than four fifths.

Nitrogen oxides are formed during the combustion of atmospheric nitrogen and nitrogen contained in fossil fuel. The latter plays a most important role especially during the burning of a pulverized solid fuel. It was found in the late 1940s by the Soviet scientists Ya. B. Zeldovich, P. Ya. Sadovnikov and L.A. Frank-Kamenetskii that the rate of creation of the so-called atmospheric "thermal" nitrogen oxides was a function of temperature and oxygen concentration in a furnace.

Later, the "fast" nitrogen oxides synthesized from atmospheric nitrogen at the front of hydrocarbon flame, and "fuel" ones formed from nitrogen gradually released during the combustion of fossil fuel were also discovered.

The total amount of nitrogen emissions from anthropogenic processes (65×10^6 tons per year) already exceeds natural inputs (40×10^6 tons per year) [3].

The major part (sometimes up to 100%) of nitrogen oxides in pulverized-coal-fired boilers is created via oxidation of atmospheric nitrogen, their amount depending on the temperature [4].

Climatic changes may also originate from the development of the nuclear fuel cycle. The stable krypton-85 possesses a high penetrating capacity due to its chemically inert properties and atmospheric mobility. Besides, neither oceans nor the earth's surface absorb it in necessary amounts, whereas the annual activity of krypton-85 if reduced to the unit capacity of atomic plants, is 1-6 times higher than other globally significant radionuclides [5].

Krypton-85 is an additional source of ionization, which affects the atmospheric electroconductivity. Calculations show that by the year 2100 the increase in atmospheric electroconductivity owing to the activity of nuclear power plants and resulting emissions of krypton-85 will be higher by 70%. Possible climatic consequences of this forecast are not yet clarified. However, it is assumed [5] that the increase of the content of krypton-85 in the atmosphere and the resulting decrease in atmospheric resistivity may lead to a drop of electric resistivity between oceans and ionosphere, which would result in frequent thunder storms, increasing amounts of thunder clouds and variation in precipitation patterns. There may also be an increase of water spouts and tornadoes.

CHAPTER II. GREENHOUSE GASES AND CLIMATE CHANGES AS VIEWED BY SOVIET SCIENTISTS

Soviet scientists almost unanimously hold that the major cause of the observed and expected climate changes in the next few decades is due to (i) the change in the chemical content of the atmosphere, and (ii) the decreasing CO_2 absorbtive capacity of oceans and forests.

A survey of Soviet scientific media indicates that the consequences of climate changes in the world and in the USSR in particular are not yet clarified in detail. The papers emphasize that serious efforts in this direction are still to be undertaken in the future. None the less, several scientists suggest that the effect of the greenhouse phenomenon on terrestrial climate is real. First of all, this effect will manifest itself in the desertification of sub-tropical areas and part of the tropics. Further, it will result in rising of levels of oceans (over the past century an increase of 10-15 cm was observed, whereas by the middle of the twenty-first century this increase will amount to 1 m). Moreover, ice thawing in permafrost areas [1] will release huge amounts of contained methane and carbon dioxide.

The recent joint Soviet-French studies in the Antarctic demonstrated that the two warm periods 6-8 and 120-130 thousand years ago were accompanied by an increased content of carbon dioxide and methane in the atmosphere. Vice versa, the cold glacial periods were characterized by a drop (about 1.5 times) in the atmospheric concentrations of the above gases.

According to Soviet scientists the present means to forecast climatic effects due to increased concentrations of greenhouse gases are far from ideal. At the same time they hold that the expected climate change may have both positive and negative effects for the USSR. At least from the point of view of agriculture they appear to be favourable, since the duration of the warm season would be increased, which would result in higher precipitation and consequently in increased agricultural productivity. However, such climatic changes accompanied by increased precipitations would also result in devastating floods and the elevation of the level of the Caspian Sea.

According to academician G.S. Golitsyn [1;11] human activity began affecting climatic processes of our planet already at the beginning of the twentieth century. On the one hand the forest area (the major carbon dioxide sink) dramatically decreased, on the other hand CO_2 emissions grew owing to large-scale consumption of fuels. Golitsyn attributes the evolution of the global climate to global temperature variations. The major cause of these changes, according to Golitsyn, seemed to consist of short- and long-term changes in the amount of solar energy reaching the earth's surface. This in turn, is related to variations of the orbit and the angle between the axes of earth to the plane of its motion around the sun, as well as to global catastrophes like volcano eruptions, collisions with asteroids, etc.

On the other hand, according to Golitsyn, the temperature is neither the sole nor the most important climatic factor affecting mankind. Precipitation is by far more important especially with respect to agriculture. The increase in the average global temperature raises the content of atmospheric water vapours, accelerates the natural circulation of water (at least in the atmosphere) and increases the average precipitation level. This increase is equal to several per cent, as far as the territory of the USSR is concerned. However, the seasonal and regional distribution of precipitates has not yet been determined. This problem is in the focus of attention of the scientific departments of the USSR State Committee on Meteorology. The results of these investigations - according to Golitsyn - will be of particular relevance to the well-being of present and future generations.

Academician Golitsyn also estimates that the average annual temperature at the earth's surface increased by 0.5° C over the last century. If one takes into account the exceptionally warm 1980s (e.g. 1988 was notably warmer than each previous year beginning from 1856), this increase would amount to 0.7° C. There have been no such rapid rates of increase in global temperatures in the history of the globe. According to the above scientist these grave trends of global climatic evolution were proved by the two previous summer seasons. They were characterized by high (above normal) amounts of precipitations in the European part of the USSR, although the summer of 1987 was cold and that of 1988 was hot. The eastern part of the United States was subjected in 1988 to a severe drought.

In the opinion of academician Golitsyn the heat power absorbed in the lower atmospheric layers and resulting in a warming of the earth's surface may be considered as a quantitative measure of the greenhouse effect. Its additional increase due to the growth of concentration of greenhouse gases is estimated as 2.6 Watts per square metre. Forty-five per cent of this value (i.e. 1.2 W) is accounted for by CO_2, 23% by CH_4, 19% - by chlorofluorocarbons and 3% - by N_2O. In the future the variation of the chemical content of the atmosphere due to the structural development of industry, environment protection programmes and at last more precise calculations are supposed to modify this ratio.

According to Golitsyn the concentration of atmospheric CO_2 increased by 25% during this century. The content of the other greenhouse gases, primarily of methane, chlorine and fluorine also increased. The increase in concentrations was stimulated by a decreased absorption of carbon dioxide by oceans due to growing saturation of its upper layers and, partially, to the spreading of a thin oil film over a vast ocean surface. The persistence of the present trends of greenhouse gas emissions of global deforestation and the pollution of oceans may result, according to Golitsyn, in an increase of terrestrial temperature by about 1° C by 2000 and 3° C by the middle of the next century.

One should distinguish between the temperature variations of the weather and those of the climate. The change of average terrestrial temperature by only one degree dramatically alters both the weather and precipitation zones, results in the retrenchment of glaciers and rise of ocean levels. The results of analysis of the Soviet scientists indicate that without the buffer activity of the ocean, the greenhouse effect would have already resulted in a temperature increase of over 1° C during the past 50 years.

This temperature increase would mostly affect the climate in middle altitudes, its impact decreasing towards tropical areas. Warm and slushy winters would become usual in the middle of Russia. A general increase in the amount of precipitations would be observed due to an intensified evaporation from ocean and sea surfaces. The first indications would be observed in the Ukraine and Middle Asia whereas the climate of the Middle East and Africa would become dry, thereby gradually increasing the desert area. The thawing of Arctic and Antarctic ice would result in increased ocean levels and the additional water would cover significant areas of land. Calculations show that the increase in the ocean level may be 1 m by 2050.

On the whole, Golitsyn points out, the above temperature increase would be favourable for the Soviet Union, since it will protract the positive temperature season and intensify precipitates. At the same time, science has not yet established the resulting impact of human activity upon biosphere and nature, and atmosphere and ocean in particular. Also, the effects of clouds and the degree of atmospheric transparence on climate would need to be clarified.

Academician Golitsyn also suggests that if in the forthcoming century the average temperature increases by 4° C and even more, mankind presumably would be capable to cope with it and neutralize several negative trends taking into account their gradual development. However, this would involve tremendous financial investments. Mankind ought to master this situation as of now, which means that it should restrict the consumption of energy resources of fossil origin and dramatically decrease the emission of detrimental substances disturbing the natural equilibrium. There is a danger of missing this opportunity, according to academician Golitsyn.

Academician M.A. Styrikovich and his colleagues [6] also suppose that technogenic variations of the physico-chemical content of the atmosphere, primarily due to the emission of CO_2, aerosols, water vapour and minor gaseous components are the major cause of possible climate changes. Dangerous consequences for mankind in their opinion also result from the destruction of the ozone layer, variation of atmospheric electroconductivity due to a change of air ionization, and other phenomena accompanying human activity.

The above scientists assume that the rise in energy consumption (12-15×10^9 tce 29/ by 2000 and 50-80×10^9 tce by the end of the twenty-first century, i.e. 0.10-0.15% of the solar energy absorbed by the earth and 0.5-1% of solar energy absorbed by land) cannot affect the global climate, provided an even distribution of the released heat could be ensured. On the other hand the emission of technogenic heat is 2-3 orders higher than its average value which may create "heat islands". Such anomalies may result in potential climatic changes on a global scale, as evidenced by simulation experiments involving total atmospheric circulation models both in the USSR and in other countries.

The Soviet scientist I.E. Zimakov [7] points out that previously plants entirely fixed CO_2 during photosynthesis. Initially, the increased CO_2 content resulted in an intensive synthesis of biomass. Nowadays, however, in numerous industrial regions plants are incapable to process the high amounts of atmospheric CO_2. The resultant increase in the content of atmospheric CO_2 leads to the greenhouse effect.

It was emphasized at a workshop in Tomsk (USSR) in April 1988 [8] that the present increase in the concentration of atmospheric CO_2 has become menacing. A further increase of concentrations towards critical values, of pollution of oceans and of deforestation could result in an irreversible and fast warming of the earth. The authors of this forecast (V.A. Shemshuk and A.V. Klyuev) hold that the warming of the earth could be accompanied by a rise of the ocean level by up to 66 m, and by more frequent tornados and huge waves. People would survive in separate groups only in high mountain areas; they would make their living by collecting plants. A renaissance of civilization would become possible only in 600-800 years, i.e. when fertile soils would be restored [8].

V.F. Krapivin, Yu.M. Svirezhev and L.P. Volkova [3] claim that the biochemical carbon cycle significantly affects energy fluxes in the biosphere. The atmospheric carbon increases the biomass of photosynthesizing plants; it is released by decaying fossil substance thus emphasizing the role of ocean and land in the process of energy transformation.

The amount of atmospheric CO_2 and the climate affect the intensity of biological carbon assimilation and decay of fossil substance, whereas the content of CO_2 in the atmosphere and consequently climate depend on the ratio of utilized and released carbon in the above two processes. Moreover, carbon changes the earth's albedo through plants and consequently affects the flux of solar energy in the biosphere.

Various proposals are made to reduce the effect of increased atmospheric CO_2 concentrations upon the climate. Most suggest to replace fossil fuel by other sources of energy as well as to increase the forest area.

As was already mentioned, the increasing concentration of atmospheric CO_2 stems, to a certain degree, from a decrease of the absorptive capacity of the main controllers of the amount of atmospheric carbon dioxide, namely forests (due to the decrease of their areas) and oceans (because of the spreading of a thin oil film).

The Soviet scientist <u>V.I. Larin</u> indicates that the main fraction of oil polluting the ocean is introduced by washing and ballast waters from ships (23%), or rivers (28%). The rest is induced by rinsing water in ports and adjacent water systems (17%), precipitations (10%), waste waters (10%), tanker accidents and accidents of offshore oil platforms (6%), or washed into the oceans by rain (5%) and drilling on the shelf (1%) [9].

<u>A.A. Fedoryaka</u> [10] points out that <u>the future of mankind is threatened by excessive saturation of lower atmospheric layers and soil with CO_2</u>. He holds that this danger is more alarming than the ultimate consequences of the greenhouse effect. Fedoryaka draw these conclusions from the following:

CO_2 has a higher density than oxygen; a thick layer of CO_2 covers the water and continental surface impeding living processes in organisms. This according to Fedoryaka has caused the disappearance of numerous species of fauna and flora. Especially dangerous effects of the CO_2 layer are observed in calm weather in low lands, in fertile soil areas in particular. The intensity of above negative process is diminished by the earth's revolution and motion of its air, which still protects living processes on earth. But even nowadays the consequences of the presence of this atmospheric layer are manifested in large-scale phenomena. They are droughts observed in certain regions and floodings resulting in excessive saturation of soil with water and washing out of fertile layer. The major cause of these phenomena Fedoryaka attributes to the ever growing CO_2 content in the lowest atmospheric layers. He assumes that the average estimations of the concentration of CO_2 in the atmosphere do not reflect the true situation, which is essentially inhomogeneous at minimum altitudes. According to Fedoryaka a significant increase in concentrations of CO_2 characteristic for numerous areas results in the three grave effects: (i) the disappearance of living organisms in the soil, resulting in decreasing fertility; (ii) the decrease in the content of oxygen in the lowest atmospheric layers leading to deteriorated living conditions for human beings and animals due to oxygen "fasting"; and (iii) the development of the "desert effect".

Calculations made by Fedoryaka indicate that the <u>CO_2 layer adjacent to the earth's surface with a thickness 11-110 cm, whose density is almost 1.5 times higher than that of air is heated more intensively</u>. This layer and the earth's surface form a specific "pitfall" for the thermal radiation of the sun. The temperature of this layer is several degrees higher than the average air temperature. Such temperature increase intensifies water evaporation from the soil. These vapours penetrate the CO_2 layer and tend upwards with the flux of warm air (i.e. nitrogen-oxygen mixture). The density of the CO_2 layer itself almost does not vary. This process results in a gradual drying of soil, which, Fedoryaka comments, may be called a "<u>desert effect</u>". Fedoryaka attributes the severe drought of 1988 in the United States of America and a peculiar redistribution of atmospheric precipitations in various regions to the above "desert effect".

The earth's biosphere, Fedoryaka concludes, became a non-equilibrium thermodynamic system due to the creation of a low-lying CO_2 layer.

In the opinion of the Soviet scientist <u>I.I. Kuz'min</u> [5] even a minimum rate of increase of global energy consumption and the maximum development of nuclear and hydroelectric stations as well as of advanced renewable energy sources may lead to an <u>increase in atmospheric CO_2 concentrations and may result by 2030 in a rise of global temperature of no less than 1° C, by 2050 1.5° C and by 2100 2° C respectively.</u> The warming up at the high latitudes would be several times higher than that of low global regions. I.I. Kuz'min also assumes that the temperature increase would be accompanied by an intensification of the hydrological cycle. For instance, <u>a growth of the average global temperature of 3° C results in an increase of the average rates of evaporation and precipitation of 6-7%.</u> It will cause an increase in humidity and consequently the further development of the greenhouse effect. Such phenomena would result in a fast increase of global temperatures. <u>The average surface temperature compared with the pre-industrial era may increase by 2° C already by 2010. By 2050 this increase could amount to 5° C.</u>

Further, I.I. Kuz'min observes [5] that a doubling of CO_2 concentrations may result in 2.4° C increase in average surface temperature. Such an increase may be favourable for certain regions. Thus, in moderate latitudes such increase must result in growing productivity of agriculture. Vice versa, in other regions these climatic changes would have a dramatic impact upon agriculture. The major problem appears to be decreasing agricultural productivity. A 1° C increase in average global temperatures may result in a 1-3% loss of harvest. These results were obtained using the model of global development. It involved two versions. In the first simulation the harvest loss was assumed to be compensated by additional investments in agriculture ("dangerous" version). The second approach suggested the use of the same investments to decrease CO_2 emission ("safe" version). The results obtained were somewhat surprising. It was shown that the "safe" version was in fact substantially more dangerous. In this case the system's development results in a crisis related to abrupt increase in mortality rate due to the drop in the living standard, since the money would be spent for the consutruction of expensive equipment controlling the emission of CO_2 into the atmosphere. On the other hand, the "dangerous" approach in fact was rather safe.

Academician <u>V.A. Legasov</u> [5] also maintained that the increasing concentration of atmospheric CO_2 might seriously disturb the climate stability resulting in a variation of precipitations and evaporation patterns. It could also cause a retreat of the snow line, glacier thawing, destabilization of the ice cover, and a perturbation of the atmosphere-ocean circulation.

<u>The intensive anthropogenic activity</u> (urbanization, deforestation, etc.) <u>might in the opinion of academician Legasov lead to an increased capacity of the earth to reflect electromagnetic radiation which would diminish the greenhouse effect and its consequences.</u> But even if the rise in average temperatures is 1-2° C lower than the calculated one, Legasov observes, it would create such living conditions that it would be difficult for mankind to adapt to them. Moreover, these changes would be long-term ones, since the once attained level of CO_2 concentration would persist at least for several centuries. This forecast was considered by Legasov as most realistic for the present level of scientific understanding. At the same time he emphasized the <u>lack of knowledge concerning the natural carbon cycle.</u> In such conditions the studies of gas transfer between the atmosphere and the ocean (the latter being the major natural source of carbon as well as the buffer storing excessive amounts of anthropogenic CO_2) become extremely important. However, the role of the ocean in absorbing CO_2 from atmosphere still needs further clarification [5].

Academician Legasov also pointed out that <u>unrestricted development of energy production based on fossil fuel may result in irreversible global climate change</u>. He assumed that the increase in the average earth's temperature in 1-2° C by 2000 and in 3-5° C afterwards would inevitably result in shifting of agroclimatic zones.

The director of the Institute of Geography of the USSR Academy of Sciences <u>V.M. Kotlyakov</u> assumes [12] that the rate of increase of anthropogenic effects on the environment became so high during the past 10-15 years that <u>the ecological dangers should be considered as compatible with the danger of nuclear war</u>. He holds that now we have to tackle the problem of environmental survival created by ourselves.

The Soviet scientists <u>E.P. Zimin and I.G. Tikhonova</u> [13] consider an increase of atmospheric CO_2 concentrations as a possible stimulation of the photosynthetic activity of plants. In their opinion <u>it is too early now to discuss serious ecological changes of atmosphere related to specific energy technologies</u>.

The burning of each carbon atom necessitates the use of two oxygen atoms. It is expected that in 2000 18×10^9 tons of oxygen will be consumed. The annual atmospheric influx of free oxygen due to photosynthesis can be estimated between 20 to 240×10^9 tons. Other calculations [14] suggest that <u>by 2010 a criticial ratio between oxygen consumption and production may be attained</u> provided the further energy development is based upon fossil fuels.

CHAPTER III. PRESENT STATUS OF THE SOVIET ENERGY INDUSTRY AND CO_2 EMISSIONS

The Soviet Union is one of the largest producers of primary energy resources and second (after the United States of America) consumer of energy.

Presently, <u>the Soviet Union produces more than one fifth (22%) of the world's primary commercial energy resources</u>, whereas the amount of natural gas, i.e. the fossil fuel with the least release of CO_2 accounts for one half of total production. The coal mining industry of the USSR (coal releasing the highest value of CO_2 per unit of heat) produces 1/6 of total world coal production.

The total production of primary energy resources in 1990 in the the USSR was estimated as $2,640 \times 10^6$ tce, 39% higher than in 1980. The major increase in the production of primary energy resources was accounted for by natural gas (71%) and nuclear and hydroelectric stations, whose total production in the present decade increased by nearly 90%. This will result in a considerable structural change in the production of primary energy sources by 1990 compared to 1980 [17].

Percentage of the total production of energy sources

	1980	1990
Oil and condensate	44	35
Gas	26	38
Solid fuel	25	19
Total amount of fossil fuel	95	92
Other primary energy sources	5	8

The leading role of ecologically advanced energy sources ensures in energy production a positive restructurizaton of the energy balance, and a decrease of the emission of atmospheric CO_2. The burning of all fossil fuel produced in the USSR would result in an atmospheric input of the carbon (as a constituent of CO_2) of roughly $1,280 \times 10^6$ tons, which is 18% higher than in 1980 ($1,030 \times 10^6$ tons). In other words, the growth rate of emission of atmospheric CO_2 resulting from the burning of the entirety of the fossil fuels produced is 80% lower than the rate of growth of total primary energy production.

If we take into account that the total primary energy production of the other European member countries of the Council of Mutual Economic Assistance (CMEA) (excluding the USSR) is estimated to be 500×10^6 tce in 1990 and that three fourths of this amount is coal then the total emission of atmospheric CO_2 by these countries may be estimated at 300×10^6 tons (with respect to carbon). Consequently, the total amount of carbon contained in fossil fuel produced in CMEA countries (including the USSR) which will be emitted into the atmosphere in 1990 can be estimated at $1,520 \times 10^6$ tons.

However, considering "contributions" of various countries in the greenhouse effect it is worth discussing not the production of primary energy resources but their consumption. This is especially important when comparing the contribution of the major producers and consumers of primary energy resources, i.e. the USSR and United States, account for more than half of the total amount of consumed energy [15].

The structure of primary energy consumption as a whole and of fossil fuels in particular with respect to CO_2 emission favourably distinguishes the USSR from the United States. This is mainly due to the fact that the share of the ecologically most clean fuel - natural gas - is 1.7 times higher (about 38% in 1990 against 26% in 1980) in the USSR than in United States (22% and 27% respectively). Thermal electric plants in the USSR, the main consumers of fossil fuel, on the average utilize 10-12% less of fossil fuels per kWh due to the combined production of electricity and heat. The above data, along with the fact that the USSR is considerably behind the United States in the absolute consumption of fossil fuels, suggest a lower contribution of the energy industry of the USSR on the greenhouse effect, compared with the United States.

The Soviet Union is one of the major exporters of energy. The net exports of energy (primary and derived ones) were 324 x 10^6 tce in 1980. Preliminary estimations suggest that in 1990 this value will increase up to 430 x 10^6 tce or almost in 1/3. This increase is mainly accounted for by the ecologically most clean energy carriers: natural gas and electricity. Export of natural gas from the USSR to CMEA countries and West Europe over the five years 1981-1985 increased by 29%. The respective figure for electricity is 51%, whereas net coal exports practically remained unchanged. In the present five-year period the structural trends in energy exports will generally continue. Hence, positive developments resulting in a decline of CO_2 emissions dominate both the domestic use and international energy trade in the USSR.

If we take into account that the burning of natural gas (on a tce basis) generates smaller amounts of nitrogen oxides (at electric plants 2.6 times less: the respective average figures are - 15.1 kg/tce. at steam electric power stations (SEPS) operating on coal and 5.8 kg/tce. at SEPS using gas), we conclude that the structural change both in domestic energy balance and in international energy trade in the USSR in the present decade should be considered favourable with respect to N_2O.

In the middle of 1989 the Director of the <u>Institute of Energy Research of the USSR Academy of Sciences and Corresponding Member of the Academy of Sciences A.A. Makarov together with I.A. Bashmakov</u> completed the study "USSR: Strategic Energy Development with Minimal Emission of Greenhouse Gases". The study indicates that <u>in 1990 1,020 x 10^6 tons of carbon will be emitted into the atmosphere due to the burning of fossil fuels in the USSR.</u> Thirty-six per cent of this amount is attributed to electric power stations, 32% - to industry, building industry and agriculture, 15% - to the residential and service sectors, 10% - to transport and 7% - to the other plants of the energy industry except the energy producing plants. Makarov and Bashmakov show that 38% of the total amount of emitted carbon originated from solid fuels, 33% - from oil and the rest (29%) - from natural gas. To facilitate the comparison we shall point out that the share of solid fuels in the total consumption of fossil fuel in 1990 will approximate 25%, of oil 34% and of gas (41%). Nearly 78% of the total greenhouse effect in the USSR originated from CO_2 emissions as a result of the burning of fossil fuels, according to the Institute of Energy Research of the USSR Academy of Sciences and the State Committee on Science and Technology of the USSR.

The Soviet scientific literature also contains other estimations. According to the <u>Fedorov Institute of Applied Geophysics</u>, 3,714 x 10^6 tons of CO_2 containing 1,014 x 10^6 of carbon were emitted into the atmosphere in 1987 due to combustion of fossil fuel. In the United States the respective figures were 4,485 x 10^6 tons (CO_2) and 1,224 x 10^6 tons (carbon). The above values do not include anthropogenic carbon dioxide created by other kinds of industrial activity (during energy production), forest fires, etc.

The results obtained by academician <u>G.S. Golitsyn</u> [1] indicate that 22 x 10^9 tons of carbon dioxide were emitted into the earth's atmosphere in 1987 (due to burning of coal: 45%, of oil: 40% and gas: 15%. On a coal equivalent basis emissions from oil were lower by 15%, and of gas by 43%. The USSR was responsible for 19% (nearly 4.2 x 10^9 tons of CO_2 or 1.3 x 10^9 tons of carbon), the United States for 23%, West Europe for 13.5%, China for 8.7%, the EEC countries for 7%, the other countries for 28% of the above 22 x 10^9 tons of CO_2. Estimations made by the Institute of Energy Research of the USSR Academy of Sciences and the USSR State Committee on Science and Technology <u>attribute to the USSR 14% of the total amount of atmospheric CO_2</u> emission. According to scientists of the above institute the higher rate of CO_2 emission with respect to national income (12.7% of the world's national income) can be explained by a higher degree of the energy intensity of the USSR economy, which is approximately 30% higher than in the United States, 78% higher than in Western Europe and 27% - higher than the world average in 1990.

According to available estimates the total consumption of primary energy in CMEA countries (excluding the USSR) will amount to approximately 660 x 10^6 tce in 1990. The consumption of fossil fuel will correspond to 95% of this amount (57% - coal, natural gas - 20% and oil - 20%). Taking into account this structure of energy consumption we can estimate the total emission of atmospheric CO_2 in the above countries in 1990 at 400 x 10^6 tons (carbon). Consequently, the total emission of <u>atmospheric CO_2 in 1990 of the European CMEA-countries (including the USSR) will be about 1,400 x 10^6 tons</u>.

There were (and still are) numerous problems facing Soviet legislation in the field of rational energy consumption, including also an underdeveloped price system for energy, economic restructuring and a bureaucratic approach to the problems of energy supply. These factors along with an unfavourable position of the Soviet engineering industry with respect to advanced technology in the field of energy efficiency are the main reasons of the high energy intensity of the USSR economy and consequently of excessive CO_2 emissons.

The higher degree of energy intensity leads to a high carbon intensity. According to calculation of the above-mentioned scientists Makarov and Bashmakov, the carbon intensity is 24% higher than in the United States, two times higher than in Western Europe and 17% higher than the world average. A comparison of the above indices with the energy intensity of national income for the USSR and the United States, however, reveals that this difference is not significant. The difference can be explained by the above-mentioned structural difference of the fuel pattern of energy consumption: the consumption of natural gas with a low carbon content in the USSR is higher. Calculations of Makarov and Bashmakov also suggest that the specific emission of CO_2 into the atmosphere related to the total value of energy utilized in the USSR is several per cent lower than in the United States and the world average. In other words, the structure of consuming branches of energy in the USSR with respect to CO_2 emission is more efficient than in the United States.

The analysis of Makarov and Bashmakov indicates that each ton of domestically consumed fuel results in the emission of 430 kg of carbon, 13% less than in the United States, the same amount as in West Europe and 19% less than the world average.

It is practically impossible to estimate the total emission of CH_4 in the USSR. Nevertheless, Makarov and Bashmakov claim that <u>in 1990 this gas will be responsible for 14% of the greenhouse effect produced by the consumption of fossil fuel in the energy sector</u>. On a carbon basis this amounts to 180 x 10^6 tons of the total amount of 1,300 x 10^6 tons of all greenhouse gases. Other calculations indicate that the role of CH_4 in the total greenhouse effect is underestimated since it does not take into account all major sources of atmospheric emissions of the above gas.

The total amount of nitrogen emitted into the atmosphere during the burning of fossil fuel in the USSR in 1990, according to the above scientists, amounts to 2.2 x 10^6 tons, of which more than half from electric power plants. Among the fossil fuels solid fuels account for 45%, oil for 13%, gas and other types of fuel for 37% of total nitrogen emissions. On a carbon basis, the emissions are approximately 110 x 10^6 tons, of which 60 x 10^6 tons or 54%.

A.A. Makarov and I.A. Bashmakov hold that 1,300 x 10^6 tons of the carbon equivalent emitted into the atmosphere from the USSR territory are due to the utilization of fossil fuels in the energy sector, of which one half in power-generation, one third in industry, buildings and agriculture and the remainder in the communal service sector and in transport.

The calculations made in the National Research Institute of Complex Fuel-Energy Problems of the State Planning Committee of the USSR indicate that the approximate source structure of air pollution was as follows (the data was collected in 1981-1985): transport – 32%, production of electricity and heat – 27%, fuel producing industry – 11%, other branches of industry – 30%.

In section 2 of the present report we have already emphasized the role of deforestation in increasing the concentration of atmospheric CO_2 and consequently in developing the greenhouse effect. The forest area of the USSR amounts to 814 x 10^6 hectares i.e. 36.6% of the USSR territory whereas the total capacity of plantings is 86 x 10^9 m^3. Approximately 3.8 x 10^6 hectares of wood are annually cut down and 290 x 10^6 m^3 of wood are annually harvested for various branches of the USSR industry. In order to conserve the forest area the annual amount of planted woods in the USSR must be equal to 2.2 x 10^6 hectares over the period 1981-1990. Thus, we may conclude that the processing of wood in the USSR cannot significantly affect their CO_2 absorbing capacity during the next two to three decades.

It is not possible to estimate the rate of atmospheric CO_2 emission in the USSR caused by the restructuration of agriculture and by the increased utilization of synthetic fertilizers in particular. Several assessments indicate that at present between 10 and 30% of the total anthropogenic emission of CO_2 can be attributed to agriculture.

Estimates of the annual global emission of greenhouse gases due to deforestation, changing agriculture, burning of agricultural waste products and the use of wood as a fuel vary from two to three times. Such estimations were not made from the Soviet Union. It is evident, however, that the above activities result in the emission of CH_4, N_2O, NO and CO along with CO_2.

A high amount of CH_4 is also generated by cattle breeding. According to some estimates 10-20% of the total amount of CH_4 (whose annual emission is about 400-600 x 10^6 tons) is produced by cattle. The respective figures for the Soviet Union are not available.

CHAPTER IV.   FUTURE ENERGY DEVELOPMENT AND THE CO_2 ISSUE

Forecasts suggest that the total production of primary energy would increase until 2000 by nearly 18% with respect to 1990. The subsequent decade would bring another 10% rise. On the whole the next 20 years would be characterized, according to the author, by an increase of 740 x 10^6 tce primary energy production by 740 x 10^6 tce, 60% of which would be accounted for by rising production of fossil fuel. The resulting increase in CO_2 emissions during 1991-2010 is 15%, nearly two times less than the growth rate of primary energy production. This slow-down can be explained by two factors.

The first factor is <u>the increase in the share of energy carriers of non-organic origin</u> (nuclear and hydroenergy, renewable sources of energy, etc.) in the total production of primary energy resources from 6% in 1990 to 13% in 2010. This results in a growing production of energy sources, whose utilization does not initiate CO_2.

The second factor is related to an <u>increasing role of natural gas in the structure of total production of fossil fuel</u> (from 43% in 1990 to approximately 51% in 2010).

If we take into account that by that time the USSR will export a significant amount of fossil fuels, the ratio between the growth rate of primary energy consumption (estimated in the above study of Makarov and Bashmakov as 2,700 x 10^6 tce in 2005 and 32,600 x 10^6 tce in 2020) and the rate of atmospheric CO_2 emissions will be altered.

We have to point out that if the <u>overall energy conservation policy</u> would not strengthen after 1990, the domestic primary energy needs of the USSR, according to the author's calculations, would have been higher by 1.3-1.5 x 10^9 tce in 2005 and by almost 3 x 10^9 tce in 2020. The major part (50-60%) of the planned conservation of primary energy should be ensured by the <u>structural modernization of the economy as a whole and its branches. One third of the expected economy of energy resources would be due to scientific and technical progress</u> leading to higher energy efficiency and advanced management.

We should also take into account <u>the substitution of deficient types of fossil fuel</u> by energy carriers of non-fossil origin. Preliminary calculations indicate that the total amount of this substitution will increase from 150 x 10^6 tce in 1990 by more than two times in 2005. It will be doubled once more in 2020.

Therefore, it is evident that the growth rate of consumption of primary energy resources will be slower than that of creation and emission of CO_2 in atmosphere.

Against the background of increasing primary energy consumption by 30% in 2005 and 21% in 2020 with respect to 1990 the growing emission of CO_2 (measured as carbon) can be estimated as 22-24% and 8-13% respectively. The total emission of all greenhouse gases (on a carbon basis) will increase by 24-27% in 2020 and 9-14% in 2005 compared with 1990. These results were also obtained by Makarov and Bashmakov. Consequently, in agreement with the above

scientists, it can be said that a modification <u>of the structure of domestic</u>
<u>consumption in favour of a reduction of the share of energy sources with high</u>
<u>carbon content and of an energy conservation policy leading to a substantial</u>
<u>increase in energy efficiency, is capable of providing a considerable decrease</u>
<u>of CO_2 emissions in the USSR</u>. This conclusion is quite evident.

According to calculations made in the Institute of Energy Research of the
USSR Academy of Sciences and the USSR State Committee on Science and
Technology, the total amount of carbon emitted as CO_2 as a result of burning
fossil fuel will amount to 1,260 x 10^6 tons in 2005 and 1,420 x 10^6 tons
in 2020, whereas the total emissions of all greenhouse gases (on a carbon
basis) will be 1,650 and 1,880 x 10^6 tons respectively (see the table).

The author has undertaken calculations for the European countries members
of the Council for Mutual Economic Assistance (excluding the USSR). These
suggest that primary energy consumption until 2005 may increase by
150 x 10^6 tce compared to 1990. By 2020 this increase may amount to
190 x 10^6 tce. On the whole in a 30-year period the annual consumption of
primary energy in the member countries of the Council for Mutual Economic
Assistance (excluding the USSR) will rise by 340 x 10^6 tce, two fifths of
which from non-fossil energy sources. Bearing this in mind and assuming a
possible minor decrease in the share of natural gas characterized by a low
carbon content we may deduce that the additional air pollution due to the
burning of fossil fuel in these countries will be equal to 70 x 10^6 tons (on
a carbon basis) in 2020 if compared with 1990.

However, all above calculations did not involve the potential of
scientific and technological progress in areas dealing with the reduction of
the greenhouse effect. They also do not take into account possible
legislative development in the sphere of energy consumption and environmental
protection as well as in the modernization of industrial mechanism in the
Soviet economy.

In this case novel technologies may for instance improve the preparation
of fossil fuels prior to burning via the extraction of carbon and removing
CO_2 and other greenhouse gases from exhaust gases.

Assuming that in the USSR the highest rise in CO_2 emissions will occur
in the utilization of fossil fuel for power-generation it is worth considering
this industrial branch separately with respect to its influence upon the
greenhouse effect.

The growth of the CO_2 content in the atmosphere caused by electric
power stations is 1.8 times higher than in all branches of industry.

Thermal power plants utilizing fossil fuel form the basis (at least up to
2010) of Soviet <u>electroenergetics</u>. Their total capacity and electricity
production in 2010 according to the author's estimates will exceed the
prospective level of 1990 by approximately 1.4 times, whereas the respective
increase in burning fossil fuels will be 1.3 times. A certain slowing down in
the rise of the rate of fuel consumption with respect to that of energy
production will be ensured by a decreasing average specific fuel consumption.
This will be the result of the construction of new, more efficient electric
power plants including combined-cycle stations characterized by higher
efficiency if compared to conventional steam power plants, further development

Table

Emission of greenhouse gases (in carbon equivalent) according to one of the scenarios of energy development in the USSR 1/ in 2020 in comparison to the expected level of 1990 in accordance with estimations made by Makarov and Bashmakov (The Institute of Energy Research of the Academy of Science and USSR State Committee on Science and Technology)

Directions of fossil fuels consumption	1990				2020				2020 with respect to 1990		
	CO_2 10^6 t	Other green- house gases, 10^6 t	Total 10^6 t	Total %	CO_2 10^6 t	Other green- house gases, 10^6 t	Total 10^6 t	Total %	CO_2 10^6 t	Other green- house gases, 10^6 t	Total 10^6 t
Electric power industry	364	136	500	38.5	620	220	840	44.7	+256	+84	+340
Fossil fuels production industries	74	31	105	8.1	100	60	160	8.5	+26	+29	+55
Other branches of industry, construction and agriculture	327	85	412	31.8	350	120	470	25.0	+23	+35	+58
Transport	106	4	110	8.5	170	10	180	9.6	+64	+6	+70
Residential and service sector	149	21	170	13.1	180	50	230	12.2	+31	+29	+60
Total	1 020	277	1 297	100.0	1 420	460	1 880	100.0	+400	+183	+583

1/ Consumption of primary energy resources in the USSR according to this scenario of energy development will be 3,260 x 10^6 tce, of which 2,220 x 10^6 are the energy delivered to final consumers, including industry - 1,030, transport - 330, residential and service sector - 500 x 10^6 tce. Consumption of fossil fuels for non-energy needs - 300 x 10^6 tce.

of efficient combined production of electricity and heat, improvement of conventional thermal electric power stations, their updating and the closing down of inefficient, worn-out power plants.

The rate of increase in detrimental emissions will be even slower. We should mention that the total amount of hazardous substances (excluding CO_2) emitted into the atmosphere by thermal power plants in the USSR decreased in the last decade by 20%.

Discussing the emission of CO_2 to the atmosphere by steam electric power plants, we should also mention the positive evolution of the structure of the fuel balance of power plants due to the increasing use of natural gas. Fifty power plants were modernized recently in such a way that they operate on natural gas instead of coal. The comparative fuel balance of thermal power stations in 1987 with respect to 1980 (as a percentage of the total use of fossil fuel) was as follows:

	1987	1980
Gaseous fuel	45.5	24.6
Solid fuel	32.8	40.3
Liquid fuel	21.7	35.1
Total amount	100	100

This restructuration of the fuel balance of thermal power plants against the background of increasing production (by 21.3% in 1987 if compared with 1980) led to the rise of emissions by only 12.9%.

According to estimates of the Institute of Energy Research of the USSR Academy of Sciences and the USSR State Committee on Science and Technology, in 1990 the carbon emissions of power plants in the USSR will amount to 360×10^6 tons, 46% originating from burning solid fuel, 20% from burning of liquid fuel and 34% from burning of gaseous fuel.

Scientific studies and developments aiming at the protection of the atmosphere from detrimental emissions from power plants involve numerous research and engineering organizations. Among them are the Krzhizhanovskij Energy Institute, the Institute of High Temperatures, Polzunov Central Institute of Boilers and Turbines, the Institute of Thermal Physics of the Siberian Division of the USSR Academy of Sciences, Dzerzhinskij All-Union Institute of Thermoengineering, the Moscow Energy Institute, etc.

The following developments should be mentioned in connection with pollution of atmosphere by greenhouse gases.

They were recently implemented in the Soviet power plants or their implementation is under way. Methods of suppression of nitrogen oxides during burning of fuel in gas-oil boilers; successive burning, introduction of water and vapour into the burning zone, gas recirculation in the flame jet; successive burning in solid fuel boilers of pre-heated pulverized coal in specially designed burners at high concentration; implementation of electron beams to purify exhaust gases of thermal power plants from nitrogen oxides; the use of electron accelerators to purify smoke of power plant, etc. are among the most advanced techniques.

The USSR scientific bodies use various approaches for the reduction of greenhouse gas emissions: (i) catalytic removal of nitrogen oxides from flue gases using ammonia and novel catalysts; (ii) analysis of ion-molecular and chemical kinetic processes accompanying radiation-chemical effects upon the flue gases during purification from nitrogen oxides resulting in development of a model accounting for a purification process and optimation of operating parameters; (iii) theoretical and experimental studies of burning and gasifying fossil fuel followed by the elaboration of two- and three-dimensional numerical models of homogeneous and heterogeneous burning involving the kinetics of creation of nitrogen oxides; (iv) development of novel technologies of low temperature, ecologically clean burning of solid fuels in fluidized bed boilers; (v) the study of the process of solid fuel burning in circulating fluidized bed boilers including the use of air-jet pre-furnaces; (vi) the search for new technical decisions to develop ecologically clean power heating boilers (less than 3 MW) with fluidized bed furnace; (vii) upgrading of the ozone-ammonia method to remove nitrogen oxides from smoke gases.

Presently, the concept of the development of the USSR thermal energy sector is being revised which will result in the development of an unambiguous strategy for the production of ecologically clean thermal energy-producing equipment. This also involves the updating of methodological aspects dealing with the technical-economic basis for the creation of an ecologically safe energy system as well as for the environmentally-sound retrofitting at existing thermal power plants.

We should like to emphasize the beginning realization of a Government-sponsored scientific and technological programme on "Ecologically Clean Steam Power Plant"; this programme aims at elaborating absolutely new technologies of power production based upon four promising kinds of solid fuel, thereby ensuring a tremendous decrease in atmospheric pollution.

The above arguments lead us to conclude that both the studies and practical implementations of ecologically clean steam power plants mainly focused on the reduction of emissions of ash, nitrogen oxides and sulfuric anhydride. Studies concerning CO_2 suppression or the purification of flue gases from carbon dioxide barely exceed the stage of preliminary research.

Several Soviet scientists hold [27] that building the most efficient purification works cannot solve the problem of protecting the environment. The true efforts in this area should not be aimed at the development of perfect CO_2 recovery techniques but against the necessity of building them.

The major approach to this problem, at least at the present stage of scientific and technological development, when there are no efficient means to remove CO_2 from flue gases of stationary and mobile power plants and industrial factories, consists of a reduction of the total amount of emissions thanks to (i) increased efficiency of utilization of energy resources and consequently the reduction of the energy intensity of the final product and (ii) upgrading the energy balance structure by decreasing the share of fossil fuels and increasing the share of energy of non-organic origin (nuclear, solar, hydraulic, wind and geothermal).

We should also mention other means of direct or indirect reduction of greenhouse gas emissions which are under way now, or whose implementation is about to start in the USSR:

(i) modernization of the national economy providing increased efficiency, a substantial rise of commodity production, and the reduction of the share of production with high resource and energy intensity;

(ii) a complete updating of economic mechanisms, a revision of the property concept and the development of an adequate price setting system. These items should minimize the waste of raw materials, fuel and energy and provide for a dramatic reduction of the resource and energy intensity of industry;

(iii) creation of conditions (convertible currency, joint ventures, policy of glasnost) for further incorporation of the Soviet Union into the world's economy;

(iv) high investments ensuring the safety of operating atomic power plants, design and construction of a new generation of atomic power plants, development of non-conventional renewable sources of energy. These measures would gain public opinion and societal institutions for nuclear power, facilitate a significant increase of the share of electric power of nuclear origin in total energy consumption, and promote a proportional rise of ecologically clean non-conventional sources of energy.

References

1. G. Golitsyn, "Climate and Economic Priorities", Kommunist, 1986, No. 6, pp. 97-105 (in Russian)

2. I.L. Karol', "Why Tanning Becomes Unpopular", Energiya: ekonomika-tekhnika-ekologiya, 1988, No. 11, pp. 9-13 (in Russian)

3. V.F. Krapivin, D.M. Svirezhev, L.P. Volkova, "Coevolution of Man and Nature", Energiya: ekonomika-tekhnika-ekologiya, 1984, No. 10, pp. 37-43 (in Russian)

4. V.R. Kotler, "Small Big Victory". Energiya: ekonomika-tekhnika-ekologiya, 1987, No. 3, pp. 46-49 (in Russian)

5. V.A. Legasov, I.I. Kuzmin, "Energetics and Planet's Climate", Energiya: ekonomika-tekhnika-ekologiya, 1988, No. 1, pp. 18-23 (in Russian)

6. M.A. Styrikovich, Yu.V. Sinyak, N.N. Chizhov, "Energetics - Ecology - Climate", Proceedings of the All-Union Research Institute of Systematic Studies, 1988, No. 5, pp. 12-18 (in Russian)

7. I.E. Zimakov, "Thinking of Tomorrow", Energiya: ekonomika-tekhnika-ekologiya, 1985, No. 4, pp. 2-8 (in Russian)

8. V.A. Shemchuk, A.V. Klyuev, "Ecologoclimatic Crisis", Proceedings of Interscientific School Seminar, Tomsk, 18-24 April 1988, pp. 103-105 (in Russian)

9. V.I. Larin, "Anxious Grumble of Ocean", Energiya: ekonomika-tekhnika-ekologiya, 1987, No. 1, pp. 36-42 (in Russian)

10. A.A. Fedoryaka, "Another Threat ...", Energiya: ekonomika-tekhnika-ekologiya, 1989, No. 6, pp. 49-52 (in Russian)

11. G.S. Golytsyn, "We Are Facing Climatic Change", Priroda i Chelovek, 1989, No. 8, pp. 28-29 (in Russian)

12. V.M. Kotlyakov, "Earth Sciences Are In Footlights", Energiya: ekonomika-tekhnika-ekologiya, 1988, No. 10, pp. 11-14 (in Russian)

13. E.P. Zimin, I.G. Tikhonova, "Dum Spiro Spero", Energiya: ekonomika-tekhnika-ekologiya, 1987, No. 1, pp. 43-46 (in Russian)

14. "Atomic Power Plants: Risk Degree", Literaturnaya Gazeta, No. 27, 5 July 1989 (in Russian)

15. I.A. Bashmakov, A.A. Beschinskij, D.B. Vol'fberg, "Comparative Studies of Energetics Development in the USSR and USA", Izvestiya Academii Nauk SSSR, Energetica: transport, 1988, No. 4, pp. 28-38 (in Russian)

16. A Nakhutin, Moskovskaya Pravda, 8 August 1989 (in Russian)

17. "USSR Energetics in 1986-1990", M.S. Vorob'ev, Yu.K. Voskresenskij, Yu.A. Goucharov et al., Ed. A.A. Troitskij, Moscow, Energoatomizdat Publ., 1987, p. 352 (in Russian)

18. "USSR in Figures in 1988", Goskomstat USSR, Moscow, Finansy; Statistica Publ., 1989, p. 316 (in Russian)

19. "Geographical Aspects of Economy-Environment Interaction. Review of the Soviet and American Literature, the USSR Academy of Sciences and American Council of Scientific Communities", Moscow, 1987, p. 186 (in Russian)

20. Yu.A. Shkolenko, "This Fragile Planet", Moscow, Mysl' Publ., 1988, p. 140 (in Russian)

21. A.A. Koshelev, B.B. Chebonenko, "Control of Energetics Effect upon Biosphere, One of Security Conditions", Voprosy obespecheniya bezopasnosti sovremennykh sistem energetiki, Voronezh, 1986, pp. 15-21 (in Russian)

22. M.T. Dmitriev, A.P. Lestrovich, "Hygienic Estimation of Environmental Pollution by Steam Power Plants", Gigiena i Sanitariya, 1986, No. 10, pp. 44-48 (in Russian)

23. V.S. Varvarskij et al., "Ecologically - Economic Grounds of Development and Operation of Urban Objects of Steam Energetics", Teploenergetika, 1987, No. 6, p. 42-46 (in Russian)

24. I.I. Kuz'min, "Energetics and Planet's Climate" Energiya: ekonomika-tekhnika-ekologiya, 1988, No. 1, pp. 18-23 (in Russian)

25. V.A. Shirokov, "Why Air Pollution Protecting Law Does Not Work?", Energiya: ekonomika-tekhnika-ekologiya, 1988, No. 11, pp. 29-32 (in Russian)

26. Yu.A. Shkolenko, "Time to Get Wise", Energiya: ekonomika-tekhnika-ekologiya, 1987, No. 4, pp. 11-15 (in Russian)

27. L.G. Mel'nik', "In Ecological Dimension", Energiya: ekonomika-tekhnika-ekologiya, 1988, No. 3, pp. 16-21 (in Russian)

28. I.I. Al'tshuler, I.V. Dobrolyubova, "Mankind and Biosphere", Energiya: ekonomika-tekhnika-ekologiya, 1988, No. 3, pp. 16-21 (in Russian)

29. G.I. Katyushina, I.Ya. Gogolev, "Turn Off the Light While Leaving", Energiya: ekonomika-tekhnika-ekologiya, 1987, No. 7, pp. 19-22 (in Russian)

31. M.A. Styrikovich, "On Estimation of Atmosphere Quality", Teploenergetika, 1987, No. 12, pp. 24-27 (in Russian)

32. V.R. Kotler, "Legislative Regulation of Nitrogen Exhaust from Steam Power Plants", Teploenergetika, 1987, No. 11, pp. 74-75 (in Russian)

33. V.I. Mamonov, N.I. Kem, "On Regulation of Amount of Air Polluting Substances by Thermal Power Plants, Methods and Means to Control Air Pollution by Industrial Exhaust and Their Applications", Proceedings of the 2nd National Conference, Leningrad, 27-29 October 1986, Leningrad, 1988, pp. 112-118 (in Russian)

34. V.L. Shul'man Several, "Actual Problems of Air Protective Activity in Energetics", Teplotekhnika, 1988, No. 8, pp. 3-5 (in Russian)

35. D.I. Tirpak, "Potential Impact of Global Climatic Change upon Surrounding Medium", Proceedings of the 5th Soviet-American Symposium, Washington, October 1988, Leningrad, 1988, pp. 137-148 (in Russian)

36. A.N. Krenke, "Global Climatic Changes: Interaction and Relation between Natural and Anthropogenic Components", Proceedings of the National Research Institute of Systematic Studies, 1988, No. 5, pp. 66-71 (in Russian)

37. V.I. Anisimov, "Big and Small Ecological Crises - Possible Solutions, Geological Sciences and Modern Global Problems", Proceedings of the National Symposium, Zvenigorod, December 1988, Moscow, 1988, pp. 9-11 (in Russian)

38. A.O. Kokorin, L.Kh. Ostromogil'nyj, "Estimation of Possible Variation of Global Methane Balance", Problemy ekologicheskogo monitoringa i modelirovaniya ekosystem, Leningrad, 1988, No. 11, pp. 224--232 (in Russian)

39. I.Ya. Sigal, "Protection of Air During Fuel Burning", Leningrad, Nedra Publ., 1988, p. 312 (in Russian)

40. Yu.A. Izrael et al., "Complex Approach to Ecological Estimation of Air Pollution Degree", Leningrad, 1988, No. 11, pp. 10-22 (in Russian)

41. Yu.D. Maslyunkov, "Improvement of National Ecological Status", Proektirovanie i inzhenernye izyskaniya, 1988, No. 6, pp. 27-28 (in Russian)

42. A.A. Logunov, "Modern Problems of Ecological Education", Vestnik Akademii Nauk SSSR, 1988, No. 11, pp. 77-79 (in Russian)

43. V.I. Duzhkov, N.I. Malyshko, "Monitoring and Status Control of Air in Europe and Prevention of Consequences of Its Pollution", Problemy kontrolya i zashchity atmosfery ot zagryazneniya, Kiev, 1988, No. 14, pp. 78-82 (in Russian)

44. G.I. Marchuk, "Ecology and Scientist's Duty", Vestnik Akademii Nauk SSSR, 1988, No. 11, pp. 3-4 (in Russian)

45. V.M. Koropalov et al., "Complete Soviet-American Studies of Global Atmospheric Distribution of Gaseous Admixtures Affecting Climate", Proceedings of the 5th Soviet-American Symposium, Washington, October 1986, Leningrad, 1988, pp. 117-129 (in Russian)

46. "Climate Changes and Their Impact upon Surrounding Medium", Priroda i resursy, 1986, No. 1-2 (in Russian)

47. O.K. Zakharov, S.V. Romanov, "Forecast of Atmospheric Carbon Dioxide Content", Meteorologiya i gidrologiya, 1988, No. 9, pp. 72-80 (in Russian)

48. V.A. Dvojnikov, O.V. Roslyakov, "Control of Nitrogen Oxide Influx via Ammonia Introduction into Burning Products", Teploenergetika, 1989, No. 9, pp. 61-64 (in Russian)

49. A.K. Vnukov, F.A. Rozanov, "Role of Steam Power Station in Pollution of Urban Areas with Developed Oil Chemistry", Teploenergetika, 1989, No. 9, pp. 64-68 (in Russian)

50. R.B. Bajramov, A.A. Petrova, "Independent Industrial Complexes Based upon Modern Instrumentation and Technology Involving Non-Conventional Sources of Energy", Teploenergetika, 1989, No. 4, pp. 8-13 (in Russian)

THE ROLE OF NEW AND RENEWABLE SOURCES OF ENERGY

CHAPTER I.   NRSE IN WORLD ENERGY NEEDS AND SUPPLIES

1.   According to various energy growth scenarios, 30/ the world will need 50%
more energy by the beginning of the next century.

2.   The share of fossil fuels - oil, coal and natural gas - is projected to
remain at 80% for the ECE region and the world as a whole. This will cause a
further rise in pollution unless new and stronger countermeasures are taken
promptly.

3.   International co-operation will be required to achieve an environmentally
sound energy supply world-wide. New and renewable sources of energy (NRSE)
will be needed to a growing extent to contribute to meeting future demand for
clean energy.

4.   In the ECE countries, the contribution of new and renewable sources of
energy is presently in the order of 10% including hydropower. It is forecast
to grow by about 50% by the year 2020. While impressive, this growth will not
be enough to raise the share of NRSE in primary energy supply beyond the
present level. However, the levels will vary from country to country. In
some ECE countries, the share of NRSE including hydropower is forecast to be
25% or more by the end of the century.

5.   Hydropower, which is currently the major renewable source of energy, will
continue for a long time to be the mainstay of energy supply among other
NRSE. However, its share, projected to account for 65% in 2020, is expected
to decrease in the ECE region, while the shares of other NRSE are expected to
grow. Forecasts for the supply shares of other NRSE mainly solar and wind as
provided by some ECE countries are given in Table 1.

Table 1

CONTRIBUTION OF NRSE, EXCLUDING HYDROPOWER, GEOTHERMAL ENERGY,
FUELWOOD, BIOMASS AND PEAT, BUT INCLUDING SOLAR, WIND,
WAVE, TIDAL AND OTEC ENERGY, TO TOTAL PRIMARY ENERGY SUPPLY
(% OF TOTAL)

Countries	Years		
	1985	2000	2020
ECE, Total	0.20	0.70	1.3-2.3
Cyprus	1.47	2.38	
Italy	-	0.29	
Portugal	-	2.67	
Switzerland	-	0.24	n.a.
Turkey	-	2.52	
United Kingdom	-	0.88	
United States	-	0.78	
Yugoslavia	-	1.46	
NB: Nuclear energy	5.80 1/	8.40	10.8

Source: United Nations ECE Energy Data Bank: Country submissions; World
Energy Horizons 2000-2020, WEC 1989.

1/   2,9% in 1980

6.   Table 1 shows that NRSE, mainly solar and wind, would increase by 15 times in absolute terms but will cover only 1.3-2.3% of total primary energy needs in 2020, compared with 0.2% in 1985.

7.   The projected contribution of NRSE in the ECE centrally planned economy countries will be lower than the ECE average. In these countries, the share of NRSE such as solar and wind, will barely reach 1% and the total NRSE share only about 6%, unchanged from present figures.

8.   Since the projected share of 80% for fossil fuels may become unacceptable for environmental reasons and resource protection, more should be done to ensure an increasing role for NRSE already in the next 20 years.

9.   The present costs of generating energy from NRSE have been the main hindrance to the more extensive use of these resources, but the rising costs of conventional energy and the need for environmentally sustainable energy supplies, warrant a significant strengthening of the efforts to improve NRSE technologies in terms of technical and economic performance.

10.   NRSE have demonstrated that in spite of the low Government budgets for NRSE RD&D compared to other sources of energy, the development of NRSE towards marketability has made substantial progress in relatively few years.

11.   By the year 2000, RD&D of NRSE will have been conducted for less than 30 years in the member countries of ECE. This is a short time considering that it took 50 years or more until other technologies such as hydropower or nuclear energy systems were ready for the market. The relatively small contributions of NRSE should be rated as remarkable, especially since figures may be underestimated due to the deficiency of the NRSE data base.

12.   One problem with assessing and monitoring the prospects and progress of new and renewable sources of energy (NRSE) 31/ is the incompleteness of the available data and the scarcity of comparable national statistics covering types of NRSE and time periods. In most publications the data on NRSE are largely aggregated (e.g. hydro + geothermal + solar together; or nuclear + hydro + geothermal together; biomass and peat included in solid fuels; ocean energy included in hydropower). Information is also inadequate on final use of new and renewable energy (heat, electricity) and categories of consumers (industry, households, etc.). Information on NRSE is scattered: it exists in energy and electricity statistics as well as in a wide range of non-energy statistical publications. Improved statistics on NRSE are essential for monitoring progress and evaluating the effects of incentives in this field of energy.

CHAPTER II. ENVIRONMENTAL IMPLICATIONS

13.   The energy supply sector is a major contributor to the environmental degradation and the depletion of natural resources.

14.   No source of energy is free from environmental impacts, but certain energy resources are better suited for environmental protection than others. New and renewable sources of energy (NRSE) belong to the latter category.

15. The main environmental problems connected with the supply and use of energy are:

   Fossil fuels: Problems arise from effluents, acid rain and from carbon dioxide emissions. According to current IEA projections, carbon dioxide emissions will increase by 2% a year. 32/ Over the next 20 years, the increase from energy production would amount to 50%. Measures for environmental protection are costly and still insufficiently developed to cope with carbon dioxide emissions from fossil fuels and their likely effects on the climate. World-wide, 90% of primary energy is currently obtained from fossil fuels. In the United Kingdom, where 80% of the electricity is supplied from fossil fuels (oil and coal), 30% of the CO_2 emissions are attributed to power generation. Although the carbon dioxide effect is not yet confirmed, it is recognized that the risk of dangerous climate change exists. Consequently, there may be a decline in the use of fossil fuels and an increase in the costs of using them environmentally soundly.

   Nuclear energy: Within recent years, public opinion has become increasingly concerned with plant failures, accidents and the risk of future disaster as well as the unsolved problem of radioactive waste disposal. As a consequence, the future share of nuclear energy in the ECE region as a whole may be lower than expected a few years ago.

   New and renewable sources of energy: NRSE, although not entirely free from environmental impacts, are environmentally more favourable than fossil fuels and nuclear energy. The negative aspects are confined to land and water use, noise and visual intrusion. NRSE basically do not create pollution, and solar and wind power in particular are advocated for environmental reasons.

16. Growing environmental considerations are giving rise to a major restructuring of energy production world-wide, in favour of environmentally sound sources and supply systems. Environmental protection policies and controls are either in place or being introduced in many countries. In the coming decade new and strong measures will be required in response to increasing environmental concerns.

17. The research note of the present study on CO_2 concentrations and energy scenarios is intended to advance decision-making on global energy response strategies to the environmental problems created by carbon dioxide emissions from power stations. The study provides various scenarios for future energy supply and the resulting CO_2 concentrations. In the preferred scenario, it is suggested to start now a gradual adaptation of the present world energy system to environmental and climate requirements, and to buy time until lasting solutions, including CO_2 removal measures, are found. The time gained should also be used for the development and further technological improvement of NRSE. The preferred scenario to adopt now would use all energy options without excessive reliance on any one source of energy but with preference given to the least carbon emitting fossil fuels and non-fossil fuels.

18. In a recent meeting of the Intergovernmental Panel on Climate Change (IPCC), whose main focus was on response strategies to climate change, emphasis was placed on the role which NRSE could play in fossil fuel substitution.

19. These studies and initiatives envisage an increased role for NRSE in the next decade, especially for solar and wind energy systems.

CHAPTER III. NRSE IN THE ENVIRONMENT

20. Measures necessary for the protection of the environment and the conservation of natural resources include the use of environmentally benign sources of energy. Renewable sources of energy are the world's most important energy reserves. Until now, their vast potential has been barely tapped, while non-renewable fuels are being depleted.

21. ECE member countries acknowledge the role of NRSE in the development of the future world energy system. Yet, the probability of NRSE assuming a greater share in energy supplies in the next two decades is currently in doubt. The long term potential, beginning early in the next century, is assessed more favourably.

22. Some NRSE systems are already well-developed, economically viable and commercially available, particularly those operating on a smaller scale, for domestic use and at remote locations. These include: solar power for water and space heating; small wind-power systems; conventional geothermal technologies; tidal systems; and biomass combustion. Other systems and sources such as wave power, need further technical development before they become widely marketable.

23. End-use can be expected to play a greater role in the shaping of future energy systems. In the ECE region, electricity and central district heating are among the favoured forms. Grid-bound and centralized energy supplies are forecast to dominate the demand increase. NRSE will gain from this trend because sources such as hydro, solar and wind will be used mainly for electricity production. Comparative electricity costs will favour NRSE already before the next century, as shown in Table 2.

24. At present, the electricity generating cost of solar photovoltaic (PV) systems is 6 to 10 times higher than oil-fueled plants. However, the cost of power from solar installations is expected to rapidly decrease while those for conventional generating sources will increase. Whether solar PV systems will penetrate central power station applications before the end of the present century hinges largely on the trend of fossil fuel prices. In solar thermal technologies, economies of scale would result in capital cost reductions, improved efficiency and accelerated economic viability.

25. Under favourable circumstances, the cost of wind energy is already lower than that of conventional energy systems. Expected capital cost reductions will further increase the comparative advantage of wind systems and will improve the prospects for energy from other NRSE in competition with conventional sources, as shown in Table 3.

Table 2

COMPARATIVE COSTS OF ELECTRICITY FROM NRSE

($US per kWh)

	1980	1989	2000
Wind	0.25	0.07	0.04
Solar thermal	0.24	0.12	0.05
Solar photovoltaic	1.50	0.35	0.06
Conventional power generation:			
Nuclear		0.04-0.13	
Oil-fired		0.06	
Coal-fired		0.04	

Source: Renewable sources of Energy, IEA/OECD, Paris 1987; International Power Generation, September 1988; WEC 1989 Survey of Energy Resources.

Table 3

COMPARATIVE INVESTMENT REQUIREMENTS FOR ELECTRICITY GENERATION
($US per unit of electricity supply)

	$US / kW	
	1980	2000
Geothermal	2 000	2 500
Mini-hydro	2 500	3 500
Wind	3 000	less than 1 000
Solar thermal energy systems (25 kWe module)	Over 1 000	
Solar photovoltaic cells (not including systems costs)	50 000	5 000
Conventional thermal	750	1 100
Nuclear	1 500	2 250

Source: Renewable Sources of Energy, IEA/OECD, Paris 1987; misc. others.

26. Available cost data for NRSE in comparison with power from conventional and nuclear sources may be unrealistic if figures for the latter fail to fully take into account the costs of fuel, waste disposal and ultimate plant dismantling. If environmental costs were added to the calculations which for fossil and nuclear energy can be 20% or more the competitive factors in NRSE power would be more apparent.

27. Investments in NRSE energy supply systems have not yet been forthcoming on a scale which would allow optimal utilization of their potential. If support were made available for research and for infrastructural costs to the extent provided in the past for other energy sources, this would produce the economies of scale required for economic viability. In turn, industry would be encouraged to invest in the production of marketable technology and equipment.

28. In the ECE region and world-wide, electricity is expected to be the fastest growing form of energy. Since power is generated predominantly from fossil fuels, increased power production means increased pollution, unless new clean-burning technologies and the greater use of NRSE, are introduced into future planning. Depending on the vigour of public support, it is estimated that NRSE could displace at least 10% of CO_2 emissions in countries with a high fossil fuel consumption for electricity. 33/

29. Electric power corporations have an important role to play in advancing the role of NRSE and consequently contributing to the improvement of the environment. If conventional fuel sources become less acceptable for environmental reasons, or more expensive, 40% of the United Kingdom's electricity requirements – as an example – could be met by NRSE, according to environmentalists promoting NRSE in the United Kingdom. A study by the Swedish electricity authority has concluded that a combination of energy-efficiency and low-carbon electricity supply strategies will be required to achieve the international commitments made at The Toronto Conference for the Protection of the Atmosphere, June 1988, to reduce CO_2 emissions by 20% by the year 2005.

30. Apart from the main area of electricity production, there is scope for the substantial substitution of fossil fuels with NRSE for heating and cooling purposes. Large quantities of conventional fuels are used today to heat and cool buildings and to generate heat for industrial and agricultural processes.

31. The contribution of NRSE towards the improvement of the environment is dependent first of all on the availability of energy sources and of financial resources for investment in these technologies. In the ECE market economies, funds for new energy investments are less of a problem than the general reluctance to invest in NRSE. In the developing countries, financial constraints are the principal problem. Funds being scarce, they will go to the lowest-cost energy resources. These countries will find it difficult to cover the additional costs of environmental controls but would invest in clean energy systems and in NRSE, if these offered the less costly alternatives.

32. The ECE region, which produces 60% of the world's primary energy and 70% of the world's electricity has the capabilities to pioneer this clean energy effort, with NRSE as an integral part, and to assist developing countries to follow suit.

CHAPTER IV.   INCENTIVES FOR NRSE

33. Governments realize that NRSE could contribute to the aims of energy and environmental policies and support the development of NRSE technologies and their utilization. Development of NRSE for commercial marketing requires funding over a long time, which like other new technologies, has limited the involvement of private industry.

34. All Governments in the ECE region have policies and programmes for NRSE, or have included NRSE development in national economic plans or energy strategies. Within such frameworks, Government action consists of subsidies for research, development and demonstration (RD&D), the creation of an institutional infrastructure, support for marketing and the promotion of economic competitiveness. Some countries such as Denmark, have set quantitative targets for the share of NRSE to be reached by the year 2000. The experience gained by ECE countries providing incentives for NRSE over a period of more than 15 years could be usefully shared with developing countries in other parts of the world.

35. Most of the RD&D programmes for the development and utilization of NRSE in the ECE member countries started as a result of the first increase in oil prices in 1973. A steady increase has been noted up to 1977 and from 1977 to 1982 for demonstration. The latter has succeeded to a certain extent to bridge the gap from research to industrial development and commercialization of the technologies and equipment. However, the levels of Government support for NRSE have not been maintained.

36. Government policies aimed at oil substitution and fuel diversification in the OECD countries, which started at the same time as the incentives for NRSE, achieved successful results because they were internationally co-ordinated and Governments were consistent in their commitment to reach the set targets within a specific time frame. In contrast, Government support for NRSE decreased sharply after 10 years, as a result of low oil prices and budgetary constraints. This decline in Government interest discouraged industry in the development of NRSE.

37. Progress in market penetration of technologies such as hydro or nuclear power, has been dependent on Government support. It gave industry the confidence to invest in building up capabilities in a new sector. The mature technologies using conventional hydro, fossil fuels and nuclear power have been and continue to be the major beneficiaries of Government financial and policy support (see Table 4).

38. Nuclear energy, for example, has been receiving Government subsidy for 50 years. With similar incentives, NRSE could be supplying as much as 25% of ECE total demand for energy by the year 2020. With increasing environmental concerns, Governments might well reconsider the question of subsidies for NRSE.

Table 4

GOVERNMENT ENERGY R D & D EXPENDITURES
(% OF TOTAL, ROUNDED, 1988)

	Other	Oil, Gas, Coal, Nuclear (non-breeder)	Advanced Nuclear	New energy sources
Austria		6	4	16
Belgium		33	55	4
Canada		78	4	4
Germany, Federal Republic of		47	30	17
Greece		23	6	61
Italy		23	30	6
Netherlands		33	6	13
Norway		58	0.2	7
Portugal		30	0	26
Spain		27	7	34
Sweden		14	12	21
Switzerland		28	21	14
United Kingdom		27	48	5
United States		47	16	6

Source: Energy Policies and Programmes of IEA Countries, 1988 Review, IEA/OECD, Paris, 1989.

39. A massive rise of 280% in spending on environmental improvements in EEC countries over the next four-year period is predicted in a survey 34/ on investing in the "green" sectors of West European industry. Environmental spending for the 12 member States amounted to $46 bn in 1987 or 0.9% of aggregate EEC gross national product. The survey predicts that this figure will rise to between $125 bn and $175 bn by 1991, representing between 2% and 3% of the EEC's gross national product. By comparison EEC Government expenditures for RD&D in NRSE were only 0.4 bn in 1988 (Denmark, France and Ireland excluded). 35/

40. If a proportional amount of the expenditures projected for cleaning conventional energy sources was expended for the development, refinement and large-scale use of renewable sources of energy and their integration into the future energy system, the desired environmental results might be accomplished at lesser cost. At least both objectives, cleaning up old sources and developing new sources, should be given equal emphasis.

41. Because the environmental and energy problems are global in scope, the United Nations must and will be a major contributor to their solution.

CHAPTER V. RECOMMENDATIONS FOR PRIORITY ACTION

42. In the light of the previous chapters, the following basis appear imperative:

(1) To improve the statistical data base on NRSE

Because of deficiencies of the existing data base and lack of international comparability, it is difficult to assess the growth prospects and monitor progress of NRSE. The development potential of NRSE and achievements already made are believed to be underestimated because of lack of internationally comparable technical and economic performance indicators. A regional data bank for NRSE to cover installed energy generating capacity; primary energy and electricity production; end-uses; etc. should be established for the period 1980 to date.

(2) To establish regional NRSE networks, and to improve and consolidate existing ones

Lack of information about development underway and experience gained with NRSE is hampering their use and further development, especially in the developing countries. Regional and sub-regional networks should be created or consolidated for the purpose of exchanging experience, sharing research, training and marketing facilities, and promoting the development and use of NRSE.

(3) To publish a Compendium of Incentives for NRSE

Many countries have gained experience in the use of various incentives for NRSE over a period of more than 15 years. This experience could be usefully shared with other countries. A publication prepared by ECE, listing the types or forms of incentives and assessing the results achieved in a comparable format, would encourage an exchange of views and improvements in this area.

(4) To incorporate NRSE in policy-making and in the agendas of all conferences concerned with the environment

NRSE has a role of growing importance to play in energy planning and can as well contribute to the attainment of environmental goals. To fulfil these potentials, NRSE must be integrated in all environmental and energy policies and be included in all discussions relating to sustainable development and environmental protection. In particular, NRSE should be included in the agendas of international conferences on the environment.

(5) To promote trade and international industrial co-operation in NRSE

For this purpose, surveys of joint venture opportunities should be carried out and the information widely disseminated in order to facilitate international trade in NRSE equipment and the transfer of NRSE technologies. This will create larger markets, encourage standardization and mass production, reduce costs, improve the knowledge of market opportunities and build an international trade infrastructure for NRSE.

PROJECTED ENERGY DEVELOPMENTS IN THE ECE REGION TILL 2010, AND POLLUTION

INTRODUCTION

A.   PURPOSE

1.   The purpose of this note is

- to assess, to the extent possible, the emissions associated with the projected energy demand/supply developments in the market economies and centrally planned economies of the ECE region till 2010;

- to discuss remedial measures, technologies and their implications; and

- to identify issues for international co-operation.

B.   METHODOLOGY, DATA BASE

2.   The primary energy projections used in this note are those described in the base document of the present study. Where necessary they have been supplemented by IEA and USSR estimates for final energy consumption and by other data available in the ECE Energy Data Bank.

3.   Pollutants considered primarily are SO_2 and NO_x, with some reference to volatile organic compounds (VOC) and CO_2. As far as polluters are concerned, attention focuses on power stations and the transport sector; however desirable in principle, the evaluation of other conversion activities (refineries, coking plants, industrial and residential boilers) and other stages of the energy flow (production, upgrading, transmission and distribution) proved impossible for lack of data.

Chapter I: PROJECTED ENERGY DEVELOPMENTS

4.   Emissions from energy activities depend on the dynamic and pattern of energy demand and supply, and on the technologies used.

5.   Part I of the present study concludes that despite considerable gains in energy efficiency primary energy demand would continue to grow in the ECE region till 2010. In absolute terms demand would grow from about 5.1 billion toe in 1985 to between 6.1 and 6.8 billion toe. The share of fossil fuels would remain dominant (1985: about 89%; 2010: between 83 and 84%). The share of the less polluting energy sources (hydro, nuclear, solar energy and natural gas) would grow from 11% in 1985 to only 16 to 17% in 2010 - the market economies projecting a higher share (20%) than the centrally planned economies (9%). This progress of the less polluting sources is beneficial in terms of environmental protection as it saves for the ECE region and by the year 2010 about 10 to 16 million tons of SO_2, 2.6 to 2.9 million tons of NO_x, 104 to 154 million tons of dust and 2,080 to 3,080 million tons of CO_2. Despite this success in avoiding pollution the study concludes that the energy projections, while feasible and viable in the short term, would not appear ecologically "sustainable" in the longer term. Remedial measures would have to be taken or strengthened to avoid environmentally "critical" loads, or where these have already been attained to reduce them.

Chapter II: PROJECTED POLLUTION

6.   These critical loads have already been exceeded with regard to several pollutants, primarily because of energy activities. Indeed, energy conversion and use is a major, if not the most important source of pollution. In the Federal Republic of Germany in 1986, 99% of NO_x emissions, 96% of SO_2 emissions, 92% of CO_2 emissions, 88% of CO emissions, 55% of volatile organic compounds and 43% of particulates originated from energy activities. 36/

7.   The growth and pattern of the projected energy demand referred to in paragraph 5 implies the risk that the reduction of pollution by 2010 may not be sufficient and may not attain targeted levels unless further control measures were taken. Moreover, new types of pollution (CO_2 and other greenhouse gases) may gain momentum as a result of the projected energy path.

(a) SO_2

8.   In 1980 SO_2 emissions in Europe stood at 54.1 million tons, of which 54% in Eastern Europe and 46% in Western Europe. According to Government plans, SO_2 emissions would fall by 28.5% (Eastern Europe: 21%, 37/ Western Europe: 14% 38/) to a level of 44.5 million tons in 2000. Most of this reduction would occur as a result of a more rational use of energy, fuel substitution and the broad application of desulphurization techniques in the power generating sector. 39/

9.   In the European Economic Community a projected increase of energy consumption between 1985 and 2010 by 210 million toe or 23% is associated with a decrease of SO_2 emissions by 45% or 6.4 million tons, provided appropriate measures are taken. 40/ An amendment of the Clean Air Act presently before the United States Congress aims at a reduction of SO_2 emissions from power stations by half, or by 10 million tons, by 2000. 41/

(b) NO_x

10.   In the market economies, NO_x emissions are likely to decrease in 1998 by 21% in comparison with the 1985 level. It is anticipated that 11 countries will fulfil their obligations to decelerate by 30% NO_x emissions over this period, and 9 countries will stabilize NO_x emissions at the 1985 level during the same period. 39/ In the European Community NO_x emissions are likely to grow from 7.7 million tons in 1985 to 9.6 million tons in 2000, but to fall thereafter to 9.3 million tons in 2010. The proposed amendment to the United States Clean Air Act aims at a reduction of NO_x emissions from power stations by 10% and from cars by 40%. 39/

11.   In eastern Europe, NO_x emissions are likely to stabilize in spite of a significant increase in motorization. 39/ At present, in the USSR, there are only 45 cars per 1,000 inhabitants, in the United States of America there are 540, in Canada 400 and in West European countries between 200 and 400.

(c) CO_2

12. The total amount of CO_2 emissions from fossil fuel combustion in the ECE region would increase from 1.5 billion tons in 1985 to 17.1-18.8 billion tons in 2010, if CO_2 emission controls would not be widely applied. Together with the other "greenhouse" gases (methane, nitrous oxide and CFC-11 and CFC-12), these emissions would induce a significant change of the climate.

(d) CO, VOC

13. The amendment to the United States Clean Air Act referred to above envisages a reduction of CO emissions from cars in cities which do not attain certain environmental quality levels, by 18% and of volatile organic compounds by 2-5%.

(e) Critical loads

14. If generally applied the envisaged measures to reduce pollution from energy activities appear significant; whether they are "sufficient" depends on the "critical loads" which ecosystems can handle. The discussion on this concept and on its quantification is presently under way. As might be expected target loads vary between ECE member countries, for example for SO_2 and for the year 2000 between 1 and 4 gS/m^2. 42/

Chapter III: REMEDIAL POLICIES AND TECHNOLOGIES, AND RELATED IMPLICATIONS ON ENERGY COSTS, DEMAND AND AVAILABILITY

A.   Remedial policies and technologies

15. Remedial policies play a decisive role in reducing pollution. Measures applied or advocated include less energy-intensive growth patterns, enhanced energy conservation and rational use of energy, promotion of the market penetration of less polluting energy sources, setting of emissions standards and penalties, encouragement of technical research.

16. In the implementation of these policies, control techniques play an important role. In the market economies, a broad spectrum of technologies is being introduced, or developed: flue gas desulphurization (FGD), selective catalytic reduction (SCR), fluidized bed combustion (FBC), integrated gasification-combined cycle systems (IGCC), gas turbines, alternative transportation fuels. In the short term till 2000, FGD and SCR are likely to play a leading role. The share of power stations equipped with FGD would double between 1987 and 2000 in the market economies. 43/ SCR penetration would be slower in power generation which generates only 20 to 30% of total NO_x emissions; but as the Austrian example shows, SCR can significantly reduce NO_x emissions (-25%).

17. Among the east European economies, the USSR implements a large-scale programme for "Clean steam power stations" aimed at developing effective control techniques for existing and new power stations. However, there are no data available on the projected speed and scale of the introduction of these technologies in the USSR and the other east European economies.

B. Related implications on energy costs, demand and availability

18. Investments in control techniques are expensive in absolute and relative
terms, but cost-effective if compared with the costs of non-action.

19. The IEA 44/ estimates the relative costs of environmental control
techniques for a whole range of pollutants at 30 to 35% of total investments
for a new coal-based power station. Retrofitting of existing plants with FGD
would add 10 to 40% to the original capital costs. A case-study for a
coal-fired power station in the Federal Republic of Germany allows an insight
into the relative costs: out of a total incremental capital cost of 47%,
19% were directed at SO_2 control, 9% at NO_x control, 6% at the retention
of particulates, 6% for the cleaning of effluents and 7.5% to cope with
noise, waste and other nuisances. Environmental controls would add 15% to
the original investments of a modern coking plant with a throughput of
2 million tons ("Prosper" in Bottrop, Federal Republic of Germany).

20. On the basis of the IEA experience, the secretariat estimates the
absolute cost of providing all coal-fired power stations in the ECE
region with SO_2 and NO_x controls at $US 300 billion (at 1980 value).
Retrofitting of all power stations in the German Democratic Republic with FGD
has been estimated to require $US 3 billion.

21. These costs will necessarily increase the specific costs. Estimated
increases per kWh range from 2 to 3% in the case of oil or gas-fired power
stations to 18 to 24% for coal-fired power stations. The amendments to the
United States Clean Air Act referred to above are said to increase the
United States electricity bill by 2%. In the Federal Republic of Germany
desulphurization increases the costs per kWh by 2 to 3 pfennig, and
denitrification by another pfennig.

22. The above-mentioned costs refer to the traditional pollutants - SO_2,
NO_x, particulates. Removal of CO_2 is not included. Preliminary estimates
suggest increases of electricity generating costs of between 25% (IGCC) and
100% (chemical absorption process) for coal-fired power stations, and about
40% for gas-fired power stations. 45/

23. Environmental controls also impact on energy demand and plant
availability. The United States Environmental Protection Agency estimates
the energy consumption of the various FGD systems at between 0.75 and 2.7% of
gross electricity generation. In the Federal Republic of Germany scrubbers
are estimated to absorb 1 to 1.5% of gross electricity production in new
plants and 3% in old plants. 46/ In comparison with these traditional control
technologies the CO_2 recovery techniques referred to in paragraph 22 would
cause capacity losses of between 11 and 28%.

24. But there are also indirect energy requirements, embodied in the
machinery and equipment and required to handle sorbants, reducing agents,
catalysts and wastes. If for example, all coal-fired electricity generating
capacities in the ECE region in the year 2000 (790,000 MW) were equipped
with FGD, the total amount of calcium carbonate needed would attain
126 million tons per year, which need to be produced, processed, shipped
and disposed of.

Chapter IV: ISSUES FOR INTERNATIONAL CO-OPERATION

25. The present note aimed at evaluating pollution, resulting from the projected energy demand/supply scenarios up to 2010 described in Part I of the present publication, and at reviewing available or emerging remedial measures.

26. This aim has been achieved only partially, primarily as a result of data imperfections but also as a result of as yet undetermined or fragmented policies. These factors affect the international analysis of the interaction between energy developments and ecosystems and, hence the direction and focus of international co-operation in this field.

27. But there are more issues or tasks:

(a) lack of international emission inventories for the various pollutants and stages of the energy flow. Particular attention needs to be devoted to a (energy, eco-) systems approach as opposed to the present focus on selected sectors (electricity generation, transport), pollutants (SO_2, NO_x) and stages of the energy flow (conversion). It appears of paramount importance that all ECE Governments participate in related activities already undertaken within the framework of ECE (Conference of European Statisticians; Senior Advisers to ECE Governments on Environmental and Water Problems); 47/

(b) improvement of emissions projections till 2000 and beyond, taking into account the planned or possible introduction of environmental control techniques;

(c) elaboration of agreed methodologies and criteria for determining "Critical Loads" which would serve as a yardstick or minimum target for the reduction of emissions from energy activities;

(d) exchanges of views on environmental policies, procedures and instruments of relevance to energy production, conversion and use, including procedures for arbitrating conflicting objectives, data on costs and benefits of protection measures, etc.;

(e) exchanges of experience with East-West transfer of environmental control techniques for energy facilities including bilateral and multilateral financing (joint ventures); consideration of a large-scale joint effort towards the transfer of energy efficient and environmentally acceptable energy technologies;

(f) identification of co-operative research subjects, such as the removal and disposal of CO_2.

SYNOPSIS—SELECTED STUDIES ON ENERGY AND ENVIRONMENTAL ISSUES

This synopsis is intended to give the reader a concise summary presentation of the main features of a number of key international studies on energy and the environment, presented in chronological order. The organization of this paper is:

(i) A listing of key features of each study comprising: (a) purpose; (b) time horizon; (c) chapter layout; and (d) conclusions. The conclusions section will be further elaborated (where appropriate) into two main areas: (i) demand/supply considerations including energy/GDP, primary energy growth rates, shares of fuels in consumption (production), electricity penetration and trade growth and pattern; (ii) feasibility/sustainability (compatibility with reserves) and CO_2 emissions. Finally there is presented a section (e) on follow-up actions.

(ii) A brief editorial section noting the overall message of the studies considered.

SECTION I

I.   OUR COMMON FUTURE (Chapter 7 - Energy)

The World Commission on Environment and Development, 20 March 1987

A.   PURPOSE OF THE STUDY

"To propose long-term environmental strategies for achieving sustainable development by the year 2000 and beyond."

B.   TIME HORIZON - to 2030 and beyond

C.   CHAPTER LAY-OUT

Introduction

I.   Energy, economy and environment
II.   Fossil fuels - the continuing dilemma
III.   Nuclear energy-unsolved problems
IV.   Wood fuels - the vanishing resource
V.   Renewable energy - the untapped potential
VI.   Energy efficiency - maintaining the momentum
VII.   Energy conservation measures
VIII. Conclusion

D.   CONCLUSIONS ON DEMAND/SUPPLY AND FEASIBILITY/SUSTAINABILITY

"Fundamental political and institutional shifts are required to restructure investment potential in order to move along ... lower (energy-efficient) paths."

"Properly managed, efficiency measures would allow industrial nations to stabilize their primary energy consumption by the turn of the century. They would also enable developing countries to achieve higher levels of growth with much reduced levels of investment, foreign debt, and environmental damage."

"Within the next 50 years, nations have the opportunity to produce the same level of energy services with as little as half the primary supply currently consumed." This would serve to significantly reduce the global output of carbon dioxide 'without any reduction in the tempo of economic growth'."

E. FOLLOW-UP ACTION

"... new dimensions of political will and institutional co-operation (are required)".

II. ENERGY AND THE ENVIRONMENT - POLICY OVERVIEW

OECD/IEA (Paris), 1989

A. PURPOSE OF THE STUDY

The study "examines the impact on energy security of existing and proposed environmental measures and discusses policy options and instruments that might be selected to achieve the objectives of both energy and environmental policies".

B. TIME HORIZON - to 2005 and beyond

C. CHAPTER LAY-OUT

Chapter I.	Introduction
Chapter II.	Background on environmental trends in the energy sector
Chapter III.	Areas of environmental concern
Chapter IV.	Typology of environmental control
Chapter V.	Fuel cycle review of environmental control
Chapter VI.	Identification and assessment of impacts of environmental measures on energy
Chapter VII.	Assessment of implications for energy security
Chapter VIII.	A framework for energy and environmental decision-making
Chapter IX.	Policy responses
Chapter X.	Policy instruments
Chapter XI.	Identification of areas for improved policy-making
Chapter XII.	Conclusions

D. CONCLUSIONS ON DEMAND/SUPPLY AND FEASIBILITY/SUSTAINABILITY

The paper is not an exercise in projections but in identification and examination of "the range of policy measures that might be taken before the year 2005 in order to accomplish energy activities in the least environmentally damaging manner and at the least cost". The elements are conceived in a four-step process as follows:

(a) Identify energy/environment interaction;

(b) Identify potential responses - (i) greater energy efficiency, (ii) add-on pollution control technologies, (iii) fuel substitution, (iv) "clean" energy technologies and (v) other (mostly non-energy related) responses, e.g. structural changes in economic systems;

(c) Identify potential instruments - (i) information, (ii) regulation and (iii) economic instruments;

(d) Develop strategy bundle - policies and instruments.

E. FOLLOW-UP ACTION

Strengthening the IEA 1990 Programme of Work are a number of areas designed "to support better policy co-ordination and analysis of policy measures and thus to smooth negative implications for energy security, where they occur ...".

III. WORLD ENERGY HORIZONS - 2000-2020

World Energy Conference (Paris), March 1989

A. PURPOSE OF THE STUDY

(a) To present a synthesis on energy supply and demand development to the year 2020;

(b) "(To compare) supply and demand projections with estimates for reserves and resources."

B. TIME HORIZON - 1985-2020 (and beyond)

C. CHAPTER LAY-OUT

Part I, General

Chapter 1 - Background
Chapter 2 - Terms of reference
Chapter 3 - Project organization
Chapter 4 - Technical framework
Chapter 5 - Scope of the study

Part 2, World balance and North-South dynamics

Chapter 1 - World consumption
Chapter 2 - North-South dynamics
Chapter 3 - Development of world supplies
Chapter 4 - World energy production
Chapter 5 - Stresses on resources
Chapter 6 - Interregional energy exchanges
Chapter 7 - Comparison of world projections
Chapter 8 - The carbon dioxide problem

Part 3, The Industrialized countries (the North)

Chapter 1 - Objectives, scope and general assumptions
Chapter 2 - Analysis of the projections retained for 2000 and 2020
Chapter 3 - Appraisal of feasibility and sustainability of the projections

Part 4, The Developing countries (the South)

Chapter 1 - Developing Countries in the world economy
Chapter 2 - The energy outlook for the South

Part 5, Orientations

Chapter 1 - Orientations for the industrialized countries
Chapter 2 - Orientations for the developing countries
Chapter 3 - Overall orientations

D-(i) CONCLUSIONS ON DEMAND/SUPPLY

(a) Energy/GDP - for the world there is projected a decline in energy intensity from 0.525 kilograms of oil equivalent per (1985) dollar of GDP to the 0.345-0.39 range, depending on growth conditions (low to moderate);

(b) Primary Energy Growth Rates - for the world the range is from 1.45% to 2% (1985-2000) and 1.0% to 1.4% (2000-2020) depending on growth conditions (low to moderate);

(c) Share of Fuels (consumption) - for the world, the 1985 percentage distribution was solid mineral fuels (27.6), petroleum products (32.6), natural gas (18.1), hydroelectric (5.8), nuclear (4.2), new energy sources (0.3) and non-commercial energy (11.5); in 2020 under moderate economic growth the share are projected to be solid mineral fuels (30.0), petroleum product (26.2), natural gas (17.4), hydroelectric (7.7), nuclear (8.2), new (2.7) and non-commercial energy (7.8);

(d) Trade Growth - "as a share of PEC (primary energy consumption), the 1985 level of almost 12% (of energy trade among major world regions) would be maintained in 2000 but, by 2020, would have fallen to around 9%."

(ii) CONCLUSIONS ON FEASIBILITY/SUSTAINABILITY

Examining cumulative production against reserves, it is stated that for coal "in the period 1985-2020, world consumption ... might reach 95 to 105 Gtoe, as against total proven reserves of 650 Gtoe". For uranium, long term stresses between cumulative production and reserves pose "serious problems ... by the end of the first half of the next century". For natural gas, by 2020 "the proven reserves of 74 Gtoe would be up to 85 or 95% exhausted, depending on the scenario (moderate or low economic growth)". For oil, "the situation is difficult for the world as a whole".

E. FOLLOW-UP ACTION

The study World Energy Horizons was presented to the 14th Congress of the World Energy Conference in Montreal in September 1989. Follow-up action can be considered, in part due to the study presented, the work programme now under way within the WEC (now World Energy Council).

IV. ENERGY FOR A NEW CENTURY

<u>Commission of the European Communities, Directorate-General for Energy
(Brussels), 3-4 May 1990</u>

<u>Report of the "Group des Sages" to the Conference on Energy
for a New Century - the European Perspective</u>

A. PURPOSE OF THE STUDY

"To draw out the key messages from the <u>Major Themes</u> report" (of the Commission's Directorate-General for Energy presented at the World Energy Conference in Montreal in September 1989).

B. TIME HORISON - to 2010 and beyond

C. CHAPTER LAY-OUT

Foreward
Introduction
Challenges - The Emerging issues
Challenges - Energy sources
The Way Forward

D. CONCLUSIONS ON DEMAND/SUPPLY AND FEASIBILITY/SUSTAINABILITY

"Completion of the internal energy market is likely to lead to a greater cohesion in the planning of gas and electricity supplies and greater integration of electricity and gas grids".

"The balance in the contribution of indigenous production and imports is changing with imports expected to increase in Community supply."

"Coal ... will grow in importance as a world energy commodity."

"A greater use of gas in the Community is now envisaged ...".

"The difficulties associated with the expansion of nuclear capacity are not only due to acceptance of the technology ... (New designs for nuclear power units) are unlikely to play a major role unless they can be made both financially attractive and overcome the opposition to nuclear power which still exists in most Member States of the Community."

"For a number of reasons ... renewable energy sources are likely to be required to play an increasing role in the long term."

"The demand for electricity will continue to grow through the 1990s."

E. FOLLOW-UP ACTIVITIES

"For environmental issues the attainment of agreed objectives based on scientific evidence could involve the setting of targets requiring new economic instruments."

V. NUCLEAR ENERGY AND THE ENVIRONMENT

International Atomic Energy Agency (Vienna), May 1989

A. PURPOSE OF THE STUDY

"... to submit to the Board of Governors (of the IAEA) programme activities towards achieving the objectives of environmentally sound and sustainable development ..."

B. TIME HORIZON - to 2000 and beyond

C. CHAPTER LAY-OUT

Introduction
General comments
Energy development
The World Commission's main issues concerning nuclear energy
Environmental aspects of nuclear energy
Applications of nuclear techniques
Significance within the Agency's budget

D. CONCLUSIONS ON DEMAND/SUPPLY AND FEASIBILITY/SUSTAINABILITY

An assessment of general directions is given based on the report of the World Commission on Environment and Development; also studies by the IAEA/OECD, CEC, IIASA and the World Energy Conference (now Council) are considered.

E. FOLLOW-UP ACTION

The IAEA Report, citing the World Commission on Environment and Development, presents a list of items on which "international agreement must be reached".

The Agency "hopes to study how quantified environmental and health impacts (of introducing nuclear power into the electricity grids of developing countries) could be incorporated into ... methodologies" (for the economic optimization of electricity system expansion).

VI. THE BERGEN MINISTERIAL DECLARATION

(The Bergen Ministerial Declaration on Sustainable Development in the ECE Region)

Ministers from 34 countries in the ECE region and the Commissioner for the Environment of the European Community, 16 May 1990

A. PURPOSE OF THE STUDY

A declaration of principles and concerns. 48/

B. TIME HORIZON - to 2005 and beyond

C.   CHAPTER LAY-OUT (sections of the Declaration)

      I.   Common Challenges
     II.   The economics of sustainability
    III.   Sustainable energy use
     IV.   Sustainable industrial activities
      V.   Awareness raising and public participation
     VI.   The follow-up process

D.   CONCLUSIONS ON DEMAND/SUPPLY AND FEASIBILITY/SUSTAINABILITY

      "To make more extensive use of economic instruments in conjunction
with a balanced mix of regulatory approaches in order to increase
efficiency of environmental protection, of the use of natural resources
and of energy consumption."

      "To work towards a co-ordinated (international) approach to the use
of economic instruments ...".

      "To support ... programmes to increase the flow of capital and
environmentally sound technology to developing and East European
countries ... to assist receiving countries on high priority resource and
environmental management projects ...".

E.   FOLLOW-UP ACTIVITIES

      "... initiate an ECE-wide campaign 'Energy Efficiency 2000' to enhance
trade and co-operation in energy efficiency, environmentally sound
techniques and management practices to close the energy efficiency gap
between actual practice and best technologies ...".

      "... promote active and close co-operation between the ECE and relevant
multilateral organizations and institutions and between the Governments
of the ECE region on the most effective forms of future co-operation on
sustainable development."

      "... recommend the establishment or continued use of Round Tables or
committees or comparable processes to promote the integration of
environmental considerations ... (to) contribute to sustainable
development ...".

      "... invite the ECE to prepare a report on the Bergen Conference as a
contribution to the 1992 Conference on Environment and Development."

      "To introduce and update an energy labelling system and voluntary
agreements or mandatory standards ... for products and processes aimed at
improving energy efficiency of buildings and appliances."   The
Ministerial Declaration called upon the ECE "to review progress in this
regard at regular intervals."

VII. THE NOORDWIJK DECLARATION ON CLIMATE CHANGE

<u>Declaration by the Ministerial Conference on Atmospheric Pollution and Climatic Change, Noordwijk (Netherlands), 6-7 November 1989</u>

A. PURPOSE OF THE STUDY

A Declaration of principles and concerns.

B. TIME HORIZON - TO THE TWENTY-FIRST CENTURY

C. CHAPTER LAY-OUT

A general statement of concerns and principles
Carbon dioxide
Chlorofluorocarbons
Other greenhouse gases
Ministerial meeting
Funding
Research and monitoring
Climate change convention

D. CONCLUSIONS ON DEMAND/SUPPLY AND FEASIBILITY/SUSTAINABILITY

"Urges all countries ... to promote better energy conservation and efficiency and the use of environmentally sound sources, practices and technologies ...".

"Agrees that industrialized countries with, as yet, relatively low energy requirements ... may have targets (for energy) that accommodate that development."

"Agrees that developing countries endeavour to meet future targets for CO_2-emissions and sinks, with due regard to their development requirements and within the limits of their financial and technical capabilities."

"Agrees that developing countries will need to be assisted financially and technically, including assistance and training ...".

E. FOLLOW-UP ACTION

"Recommends that this (Noordwijk) declaration and the supporting papers be conveyed to the IPCC ... for further consideration and action."

"Urges all countries and relevant organizations to increase their climate change research and monitoring activities ...".

"Recommends that more research should be carried out by 1992 into the source and sinks of the greenhouse gases other than CO_2 and CFCs ...".

VIII.   INTERRELATIONSHIPS BETWEEN ENVORONMENTAL AND ENERGY POLICIES

ENERGY/AC.10/R.2/Rev.1, 15 February 1990, United Nations
Economic Commission for Europe (Geneva)

A.   PURPOSE OF THE STUDY

(a) Developing and analysing projections of energy demand and supply to 2000 and 2010, and evaluating longer-term scenarios elaborated in other fora;

(b) Analysing the feasibility and sustainability of these projections and scenarios from a long-term and region-wide standpoint;

(c) Describing policy responses as implemented or envisaged by ECE Governments;

(d) Identifying and discussing the issues relevant to the process of adaptation to environmental concerns; and

(e) Defining the role of regional energy and environmental co-operation through ECE in furthering this process of adaptation.

B.   TIME HORIZON – 1985-2000 (and beyond)

C.   CHAPTER LAY-OUT

Chapter I.   The interface between environmental and energy policies – the basic issues
Chapter II.   Trends and projections of energy developments until 2010 and beyond
Chapter III. Appraisal of the sustainability of these projections and scenarios
Chapter IV.   Adaptation of present energy developments and policies
Chapter V.   The role of regional energy and environmental co-operation, particularly through ECE and Senior Advisers to ECE Governments on Energy

D.-(i) CONCLUSIONS ON DEMAND/SUPPLY

Energy/GDP – "... a significant decline of the energy intensities by one quarter to one third" (for the ECE region).

Primary Energy Growth rates – 1.0-1.4% per annum for the ECE region as a whole during 1985-2000; for 2000-2010, 0.3-0.7% per annum.

Shares of Fuels (consumption) – "fossil fuels would fall from 89% in 1985 to 83-84% in 2010, with coal and gas maintaining or slightly improving their shares, oil losing, and nuclear power and new sources improving their relative position" (for the ECE region).

Electricity Penetration – for end use to cover 41% in 2010 vs. 32% in 1985 (for the ECE region).

Trade Growth and Policies – net imports of the ECE region are projected to rise by 29-30% over the 1985-2010 period.

(ii) CONCLUSIONS ON FEASIBILITY/SUSTAINABILITY

Reserves - sufficient to meet projected ECE demand by 2010 but with pressures on conventional oil reserves by the end of the period.

CO_2 - "whether the projected growth and pattern of fossil fuel uses in the industrialized countries would be sustainable, is a matter under consideration and research".

FOLLOW-UP ACTION

The study "identifies issues which the Senior Advisers to ECE Governments on Energy may wish to consider at (its) seventh session (in the fall of 1990)".

IX. POLICY-MAKERS SUMMARY, IPCC WORKING GROUP III

Intergovernmental Panel on Climate Change, First Draft, 1 May 1990

Draft report of Working Group III under revision following a 5-9 June 1990 meeting with a final Report to be presented at the fourth session of the (full) IPCC plenary to be held in Sundsvall, Sweden from 27 to 30 August 1990

A. PURPOSE OF THE STUDY

"(To) provide an interim assessment for policy makers. (The Draft Report) identifies the major issues which need to be considered in the development of national, regional and international climate change response strategies. The report also provides an overview of the main issues which should be considered in future international negotiations related to climate change."

B. TIME HORIZON - 1985 to 2025 and beyond

C. CHAPTER LAY-OUT

1. Introduction
2. Anthropogenic sources of greenhouse gases
3. Future greenhouse gas emissions
4. Climate change response strategies
5. Options for limiting greenhouse gas emissions
6. Greenhouse gas emission targets
7. Measures for adapting to climate change
8. Mechanisms for implementing response strategies
9. Conclusion

D. CONCLUSIONS ON DEMAND/SUPPLY AND FEASIBILITY/SUSTAINABILITY

"There is no single, quick fix technological option for limiting greenhouses emissions from energy sources. A comprehensive strategy is necessary which deals with improving efficiency on both the demand and supply sides as a priority and emphasizes technological research, development and deployment."

"factors external to the energy sector also significantly constrain
potential. These include the difficulty of - making basic changes in the
structure of economies ... (and) making fundamental changes in
attitudinal and social factors ...".

E. FOLLOW-UP ACTION

The final report and consideration of the future role of the IPCC itself
is to be taken up at the Sundsvall meeting.

SECTION II

The studies briefly outlined above vary from declarations of principles
(The Noordwijk Declaration, the Bergen Ministerial Declaration, and the (IPCC)
Policy-makers Summary) to reports aimed principally at a policy discussion.
There are a number of common elements to most studies with certain common
themes and but also differences. A few comments on these themes would seem to
be appropriate.

THE GREENHOUSE GAS PROBLEM AND ENERGY POLICY - The final meeting of the
IPCC this year in Sundsvall, Sweden will take on the task of furthering the
debate on this issue. It is recognized that emissions of carbon dioxide,
associated with fossil fuel use, is far above the rate needed to stabilize
concentrations of CO_2. The debate appears to be centred on how to cut the
rate of growth of fossil fuel use both directly and through substitution of
other fuels and encouraging energy efficiency measures. How to share costs of
a transition to a more rational world energy diet between developed and
developing countries will be a major focal point of the debate. All studies
reviewed contribute to an understanding of this issue. Perhaps the reader
should carefully note the difficulties cited in the CEC Study which underline
the problem of achieving progress if efforts are made only regionally but not
at the world level.

ENERGY GROWTH PATHS - World Energy Horizons projects energy demand growth
rates of 1.45% to 2% for 1985-2000 and a lower bandwidth of growth rates from
1.0 to 1.4% in the years 2000-2020. The upper end of this range reflects
economic growth of 3.2% in the 1985-2000 period with a 2.8% annual rate in the
2000-2020 period in the "moderate growth" scenario; for the low growth
scenario, the corresponding growth rates are 2.4% (1985-2000) and 1.8%
(2000-2020). This pattern which combines falling growth rates and increasing
efficiency also characterizes Interrelationships Between Environmental and
Energy Policies (in the ECE countries).

THE PRESSURE ON RESERVES - Present consumption rates appear to be
sustainable in the period to 2010, according to the ECE and WEC studies. But
beyond this period, there appear to be problems in matching accumulated
consumption and reserves, at least for hydrocarbons and uranium. Thus the
need to press forward with the development of both new forms of energy and
greater efficiency would appear to be a logical policy even in the absence of
environmental concerns.

THE DEVELOPMENT OF NEW ENERGY TECHNOLOGIES - <u>Energy and the Environment</u> is a prime source covering this area. If success is to be realized in achieving greater energy efficiency and pollution control, the blueprint offered by the OECD/IEA study will no doubt serve as a catalyst to further progress.

THE NUCLEAR ISSUE - At present, the future role of nuclear energy is obviously an unsettled matter in the international debate on the role of various forms of energy in satisfying future needs. The studies by the ECE, IAEA, the World Commission on Environment and Development and the CEC are all useful in understanding what the elements of the debate are.

ECONOMIC INSTRUMENTS - How to achieve a better energy balance with the use of economic instruments is a key subject taken up in a number of the studies reviewed. Perhaps the discussion in the OECD/IEA study is as useful as any in giving the reader a working background in this area.

ENERGY EFFICIENCY STANDARDS - This is a common theme noted by several studies. In particular, the Bergen Ministerial Declaration call for the ECE to review progress in this area is to be noted as well as substantial treatment of this subject in Chapter IX.2 (Greater Energy Efficiency) of <u>Energy and the Environment</u>.

THE LACUNAE - One recognizes the difficulty of addressing the energy policy problems of the next century. But that is what has to begin in a serious way if the (apparent) threat posed by the buildup of greenhouse gases is to be met. If policy-makers are to go beyond the first step of limiting greenhouse gas emissions (or even to cut them to a certain degree) but to achieve a <u>sustainable equilibrium</u> between emissions of greenhouse gases and the earth's ability to absorb these emissions, then a number of issues will have to be addressed. The technical question of what is the sustainable rate of greenhouse gas emissions is the first order of business. Then the consequences of accommodating the equilibrium rate (which is purported to be much lower than current rates - noting the work of the IPCC) are the need to both encourage energy efficiency steadily over a very long period of time, perhaps throughout the next century, and at the same time to develop new energy sources to substitute for the presently fossil-fuel-wedded system. The transition to a sustainable state might even require a full century to achieve, as is suggested in EC.AD/R.28, <u>Sustainable Development - Framework for Discussion of Selected Issues</u> (in particular, Chapter III. Fossil Fuel Use and Sustainable Development Paths - A case Study). More work clearly has to be done in this area.

Notes

1/   John S. Hoffman (Director of the Strategic Studies Staff,
United States Environmental Protection Agency) and John Bruce Wells in their
review paper, "Forests - Past and Projected Changes in Greenhouse Gases",
in The Greenhouse Effect, Climate Change and U.S. Forests, 1987, cite the
Nordhaus and Yohe work as "the most widely accepted set of projections". For
a review of other projection approaches, see Policy Options for Stabilizing
Global Climate Change, Draft Report to Congress, Volume I, United States
Environmental Protection Agency Office of Policy, Planning and Evaluation,
February 1989 (hereafter referred to as the EPA Report), Chapter I, pp. I-15
to I-32 (Studies of Future CO_2 Emissions). It should be emphasized that,
at this time, CO_2 projections models are not well-established on the basis
of long time series of data and generally agreed upon theoretical model and
therefore are subject to a degree of uncertainty, explained below for
the Nordhaus and Yohe approach used here. In the EPA Report, while
CO_2 concentrations are calculated on the basis of emission data (see
Chapter V, Atmospheric Concentration), the model of which these calculations
is not specified in the Report. While there are other approaches linking
carbon emissions and CO_2 concentration levels, uses of the Nordhaus and
Yohe technique appears to represent an advance over previous efforts where
atmosphere concentration levels are linked ONLY to past concentration
levels and current emission levels without taking into account, explicitly,
of seepage. The work of Jae Edmonds and John Reilly, in "Global Energy and
CO_2 to the Year 2050", The Energy Journal, July 1983 (pp. 21-48) may be
cited as an example of this approach. AGAIN, THERE IS NOT SUFFICIENT TIME
SERIES DATA FOR BOTH EMISSION LEVELS AND CO_2 CONCENTRATIONS TO JUDGE
(ECONOMETRICALLY) BETWEEN COMPETING MODELS OF THE EMISSION/CO_2 CONCENTRATION
LINK, at least it would seem at this point in time. (It should be noted that
a series of carbon emissions does exist from 1880-1980 and is reported in
R.M. Rotty and G. Marland, The Changing Pattern of Fossil Fuel CO_2 Emissions,
DOE/OR21400-2, United States Department of Energy, Washington, D.C., 1984.
However, this series (unlike the direct observations made at Mauna Loa
Observatory (Hawaii), are derived from estimates of fossil fuel use. In
Hoffman and Wells, op. cit., in reviewing this data, state that "the natural
losses of CO_2 presumably balanced the natural emissions, leading only to a
seasonal cycle still visible each year". Thus there are scientific grounds
for supporting the inclusion of a natural seepage factor as the basis for
long-term equilibrium of CO_2 concentration. The Nordhaus and Yohe approach,
on the other hand, specifically allows for a natural decay rate (seepage),
an approach to which the Report of the Carbon Dioxide Assessment Committee
attached particular importance. This approach allows for an equilibrium level
of CO_2 concentration for every steady state emission rate. As a first
analytical step, accordingly, the Nordhaus and Yohe approach is taken here
with the reservation that other approaches, as the field develops further, may
more exactly treat the problem under consideration. (Also see Climate Change,
op. cit., Section 1.2.2 (Future Atmospheric CO_2 Concentrations) both for
a review of projections efforts and uncertainties in establishing CO_2
concentrations with a view to taking into account the entire carbon cycle.)
It should be noted that the parameters used by Nordhaus and Yohe accord well
with the 28 years of CO_2 concentration observations taken at Mauna Loa
Observatory and the associated fossil fuel carbon emission data given in
the EPA Report, op. cit., p. I-10. Using the Yohe equation and an average
330.5 ppm concentration of CO_2 over the 1958-1986 period (ibid., p. II-8),
and employing the Yohe values for the marginal airborne fraction of carbon

remaining in the atmosphere (0.47) and a seepage factor of -.001, the implied average emission rate of carbon emissions from fossil fuels is 3.5 gigatons per year, well in accord with the (graphical) record presented in the EPA Report (quantitative data were not given). The graphical record for CO_2 emissions and concentration are reproduced here as figure I.

2/ Climate Change, op. cit., p. 23.

3/ The Nordhaus and Yohe projections do not explictly take account of non-fossil fuel sources of carbon, estimated to result in a total emission rate of perhaps 5.9 gigatons (in the EPA Report, op. cit.). The parameters (0.47 and -0.001) as they relate fossil fuel emissions to CO_2 concentrations are subject to possible error in one or both parameters. But without a time series for non-fuel emissions, it would be difficult to assign a quantitative magnitude to the bias, although the effect (as noted in Climate Change, op. cit., p. 1) is to overstate the marginal airborne coefficient of carbon remaining in the atmosphere, for a given seepage factor. This view, however, is challenged by some researchers who suggest that "as the oceans become saturated with CO_2 ... the percentage of world-wide carbon dioxide which remains in the atmosphere (from current emissions) ... will increase". (See Hoffman and Wells, op. cit., p. 22.) This latter argument might also raise questions on the time-constancy of the seepage factor as well. With these caveats duly noted, the Nordhaus and Yohe equation is used as originally presented and using only fossil-fuel related emissions of carbon (as the coefficients are derived on this basis). Put differently, built into the Nordhaus and Yohe approach are two implicit assumptions: (a) that the ratio of fossil fuel emissions to total emissions is constant (which permits the use of fossil fuel emissions rather than total emissions in their equation) and (b) that fossil fuel emissions may be approximated by multiplying each fuel source by a fixed coefficient, assuming a constant share of economic activity for directly CO_2 producing activities, namely cement production. Surely further research (and availability of more data) will make explicit these relationships.

4/ A given emission rate corresponds to the same growth of fossil fuels for a constant share of coal, oil and gas. For a constant share of oil, a higher (lower) physical growth rate of fossil fuels for the same emission rate occurs for a rising (falling) share of gas and corresponding falling (rising) share of solid fuels.

5/ However, if the marginal airborne fraction of carbon is much higher than 0.47, a lower growth rate of emissions would reach the 550 ppm level before 2030.

6/ Given the relatively higher carbon emission rate of solid fuels as compared with liquid fuels and gas, the cut in the total use of carbon-based fuels could be less if the role of gas increases relative to that of coal. The emission factor ranges cited in K. Todani, Y.M. Park and G.H. Stevens, "The Near Term Contribution of Nuclear Energy to Reducing CO_2 Emissions in OECD Countries", OECD Nuclear Energy Agency (1989) are as follows: for gas, 0.43 to 0.49 Gton/TWyr, for oil, 0.62 to 0.66 Gtons/TWyr and for coal 0.75 to 0.88 Gtons/TWyr. The data presented must be used with caution as they are based on sample estimates which reflect the characteristics of the (fuel) fields included. With this caveat in mind, assuming that the relative emission factors are reasonably accurate world averages, the weighted average

emission factor (using mid-points for the range shown) for 1985 carbon-based
fuels was about 0.66. The 1985 shares of solid fuels, liquid fuels and gas
within carbon-based fuels was 35.3, 41.6 and 23.1% respectively. A shift, for
example to a mixture of carbon-based fuel mix containing 25% solid fuel, 35%
liquid fuel and 40% gas would lower the weighted average emission factor to
about 0.61 which would reduce the CO_2 emission rate by about 7.3%. Thus the
1985 fossil fuel emission rate would have been about 5.1 gigatons rather
than 5.5 gigatons if the present fuel mix were to have been replaced by the
hypothetical one examined above. Put differently, the cut in carbon-based
fuels required for a 2.6 gigaton fossil fuel emission rate of CO_2 in 2090
would be about 200 mtoe lower with the (alternative) fuel mix in which the
role of gas is greater.

7/ Smith, op. cit., p. 44. Also see International Coal Report,
August 1989 (Greenhouse and Coal) for a discussion of greenhouse gases and
control options.

8/ The preferred scenario is based on the moderate economic
growth scenario of the World Energy Conference publication, Global Energy
Perspectives, 2000-2020 (Montreal, 1989). This scenario covers the world
as a whole, of which the ECE region accounted for 68% of primary energy demand
in 1985 and is projected to account for 56% in 2020. This reference scenario
is amended and extended to 2040 for the purposes of the preferred scenario
as follows. The growth rate of world GDP is assumed to be the same in the
2020-2040 period as in the growth rate specified in the moderate growth
scenario for 2000-2020, i.e. 2.79% per annum. For the purpose of analysing
carbon emission rates, specific emission factors were assumed to apply at the
world level (and are consistent with the 1985 total fossil fuel emission rate
for 1985 (5.5 gigatons of carbon) reported in Yohe, op. cit.). The specific
emission rates for fossil fuels used in Gtons/TWyr (Gtons/mtoe) are: 0.8702
(0.001155) coal, 0.6526 (0.0008663) oil and 0.4845 (0.0006432) gas. Using
these emission rates, the total fossil fuel emission rate was calculated and
is shown in table 5; the compound annual growth rate for emissions (on which
basis an emission path for the projection was established) was projected to
be 1.66 for 1985-2000 and 0.99 for 2000-2020. For non-fossil fuels, the total
for non-commercial energy was assumed to be the same in 2040 as in 2020 with
the remaining non-fossil energy sources (hydro, nuclear and new) in the same
relative proportions in 2040 as in 2020.

9/ This discussion relies primarily on Michael J. Scott,
James A. Edmonds, Mark A. Kellog and Robert W. Schultz, "Global Energy and
the Greenhouse Effect", Session 2.1.1 of the Fourteenth Congress of the World
Energy Conference and in Hoffman and Wells, op. cit.

10/ Ibid., pp. 10-11. Also taken into account is the impact of
non-RIGS trace gases including aerosols, water vapour and ozone. Also see
the discussion in Smith, op. cit., part 3 (Emissions of Other Greenhouse Gases)
and part 4.3 (Other Greenhouse Gases) where it is concluded that increased
concentrations of the other greenhouse gases serve to reinforce the impact
of higher CO_2 concentration levels (p. 29). This same conclusion is also
reached in Climate Change, op. cit., p. 2 (and Chapter 4 of this study).
The quantitative importance of greenhouse gases and their growth rates
are also given in H. Frassal, "Minderung des Treibhauseffektes",
Energiewirtschaftliche Tagesfragen 1989, Heft 12.

11/ REUS Technical Series 5. Dissemination of Renewable Energies in Farms and Rural Villages, FAO/UNDP, January 1988, pp.50-51.

12/ Including transportation and distribution investments for fossil fuels and electric power.

13/ Overall Economic Perspective to the Year 2000, New York 1988, United Nations, N.Y., p.12.

14/ Energy Conservation Policy, Case Study for Hungary.

15/ Including transporation and distribution investments for fossil fuels and electric power.

16/ Jose Goldemberg, et al., Energy for a sustainable world, New York 1988, p.296.

17/ Prepared by Wolf Häfele, Director-General, Nuclear Research Centre, Jülich, Federal Republic of Germany.

18/ Under the assumptions made, the carbon content CO_2 emissions in the ECE region would fall in 2000 from 4.8 gigatons (Gt) to 4.4 Gt and in 2010 from 5.1 Gt to 4.7 Gt. Global CO_2 emissions would fall in 2000 from 7.2 Gt to 6.8 Gt and in 2010 from 8.9 Gt to 8.5 Gt.

19/ Indicative Planning Figure (IPF) countries are those entitled to receive United Nations technical assistance.

20/ See Energy Efficiency in European Industry, ECE Energy Series No. 1, United Nations, New York, 1989.

21/ Energy Conservation in IEA Countries, International Energy Agency, OECD, Paris 1987.

22/ Ibid.

23/ "Oil Conservation: Permanent or Reversible? The Example of Homes in the OECD", L. Schipper and A. Ketoff, Lawrence Berkeley Laboratory, University of California, 1984.

24/ Improved Techniques for the Extraction of Primary Forms of Energy, United Nations Economic Commission for Europe, Graham and Trotman Ltd., London, 1983.

25/ Energy Conservation in IEA Countries, op. cit.

26/ Energy for a Sustainable World, J. Goldemberg, Wiley Easter Ltd., New Delhi, 1988.

27/ Indicative Planning Figure (IPF) countries are those entitled to receive United Nations technical assistance.

28/ Prepared by D.B. Volfberg, Ph.D. (Economy), Deputy Chief of the Department of Priority Directions of Progress in Science and Engineering of the Fuel-Energy Complex of the USSR State Committee on Science and Technology.

29/ Tce: tons of coal equivalent.

30/ J.R. Frisch: World Energy Horizons 2000-2020, Paris 1989.

31/ Defined by United Nations General Assembly resolution 33/148 as comprising 14 sources: hydro, solar, geothermal, wind, tidal, wave energy, thermal gradient of the ocean, biomass, fuelwood, charcoal, peat, energy from draught animals, oil shale and tar sands.

32/ The Industrial Energy Club, Energy Review Summer/Autumn 1989, pp.1-3.

33/ Ibid.

34/ Investing in a Green Europe, UBS Phillips and Drew, London, 1989.

35/ Energy Policies and Programmes of IEA Countries, 1988 Review, OECD/IEA 1989.

36/ Deutscher Bundestag, Schutz der Erdatmosphäre, Bonn, 1988, p. 485.

37/ Excluding Romania and Yugoslavia.

38/ Including Yugoslavia.

39/ EMEP/MSC-W, Note 3/89.

40/ EEC, Energy in Europe, September 1989, Annex IV.

41/ The Energy Report, 19 June 1989, pp. 446-447.

42/ EB/AIR/GE.2, Table 1.

43/ IEA, Emission Control, Paris, 1988, p. 48.

44/ IEA, Emission Control, Paris, 1988, pp. 102 and 114.

45/ K. Blok, Ch. Hendriks, W. Turkenburg: The Role of Carbon Dioxide Removal in the Reduction of the Greenhouse Effect, University of Utrecht (Netherlands), April 1989, Tables 1 and 2.

46/ IEA, op.cit., p. 130.

47/ Executive Body for the Convention on Long-range Transboundary Air Pollution.

48/ In the context of the 1990 Bergen Conference, the Report (to the Bergen Conference) of the Velen Workshop on "Energy and the Environment - Sustainable Energy Use" is to be noted in particular as it served as a key background study. The Velen Workshop Report presented (a) a summary of "present understanding of the implications of the concept of sustainable development vis-à-vis the problems associated with climate change and other energy-related environmental effects" and (b) the identification of "possible objectives, policy instruments and specific action for national governments, international and non-governmental organizations in an effort to achieve sustainable energy use and production".

PART III

THE ENERGY PROGRAMME OF UN-ECE

A.   THE IMPORTANCE OF ENERGY CO-OPERATION

1.   Energy, as a pre-condition for development and welfare, played an important role in the programme of work of ECE since its inception. Among the first Principal Subsidiary Bodies established in 1947 were a Committee on Electric Power and a Coal Committee. At present, about one quarter, or approximately 180 projects, of the Commission's five-year programme of work are related to energy.

B.   THE EVOLUTION OF CO-OPERATION SINCE 1947

2.   Whereas in the post-war period, the allocation of scarce coal and the production of electricity from coal and hydropower were in the forefront of interest, changing circumstances prompted later on an expansion and diversification of programmes. Topical issues included:

-   in the 1950s: mechanization and automation in coal mining; substitution of oil and gas for coal, by consumption sectors; trade classification of hard coal by type; assessment of Europe's hydropower potential; rural electrification; production and use of manufactured gas; internationally-comparable coal, electric power and gas statistics;

-   in the 1960s: technical progress; new management tools, such as operations research and computers; integration of nuclear power; interconnection of Europe's power and gas networks; substitution of natural gas for town gas; transport of energy, including long-distance transport and transmission; development of the energy supply infrastructure; energy modelling;

-   in the 1970s: expansion and diversification of east-west energy trade; multi-fuel integrated energy policies; long-term demand/supply projections; the interface between demographic, macro-economic and energy trends; internationally-comparable general energy statistics; reduction of air and water pollution; energy economy and efficiency; interdependence of national and regional energy systems; diversification of energy trade; new technologies and energy sources; energy data banks.

C.   PRESENT EMPHASIS

3.   The Commission's present energy activities may be divided into general and policy-oriented activities, fuel-specific activities and user-specific activities.

(a)   General and policy-oriented activities

-   Macro-economic context: overall economic perspective for the ECE region to the year 2000; interrelationship between energy prospects and long-term macro-economic trends; impact of structural changes in industry in ECE countries on the level and pattern of energy requirements of these industries; long-term demographic trends and prospects in human resources.

- Overall policies: national energy programmes, policies and prospects; energy transition policies in the region; significant new developments affecting energy demand and supply in the short, medium and long run; end-use demand analysis; optimum use of primary energy resources in final heat consumption.

- Co-operation: problems and opportunities of east-west energy trade and co-operation; east-west industrial co-operation in the field of energy; identification and analysis of energy projects suitable for international co-operation.

- Energy economy and efficiency: developments, policies and prospects; long-term impact of energy efficiency improvements; recovery and rational utilization of secondary forms of energy; gas saving and rational use; efficiency of gas utilization appliances and equipment; gas and electric heat pumps; reduction of electric power losses.

- Environmental protection: interrelationship between environmental and energy policies; environmental aspects of opencast mining and coal utilization; environmental aspects of the gas industry; electric power and the environment; impact of power stations, reduction of emissions, combined production of electricity and heat, radioactive effluence, fluidized-bed combustion, solid wastes, cooling towers, transmission lines.

(b) Fuel-specific activities

- Coal: prospects for the coal industry in the ECE region; world coal trade up to the year 2000; short-term fluctuations in the demand for and supply of solid fuels; increased use of coal in industry and district heating; technical progress in underground mining; measures to improve mine safety; technical progress in opencast mining, including environmental protection; coal research and development; coal preparation and utilization, utilization of low-calorific-value fuels; use of computers for process control and data processing; ECE classification of coals; promotion of the wider use of coal in industry and other sectors; capital formation and costs of production; coal statistics and information.

- Electric power: medium- and long-term prospects and policies for electric power; electric power generation; nuclear power, fossil fuels, hydroelectric schemes; new developments in geothermal energy; electric power systems and interconnections; electricity distribution, equipment and networks outside the major urban areas; rational use of electricity; electric heat pumps; reduction of electric power losses; electric power and the environment; electric space heating; combined production of electricity and heat; electric power statistics; glossaries.

- Nuclear energy: adaptation of nuclear power stations to network requirements; condensates and feedwater treatment and preservation; down-time; education and professional training; fast neutron reactors; review of national nuclear policies in the context of general energy policies; reduction of inconveniences during the construction of large nuclear power plants; ecological effects arising from the use of cooling water.

- <u>New and renewable sources</u>: status and prospects; comparative merits of various new and renewable sources on the basis of common methodological guidelines; storage; low-temperature solar heat application; Europe's hydroelectric potential; role of hydro, solar and combined power and heat stations in covering load; geothermal energy for electricity production and space heating; management of electrical energy on farms equipped with non-conventional energy sources; measuring radioactivity in the effluence from nuclear power stations.

- <u>Gas</u>: future role of gas in meeting energy requirments; long-term development of the LPG industry; recent developments in gas availabilities and consumption; intra- and interregional trade in gas; environmental aspects of the gas industry; gas resources; gas transport and storage, including underground storage; offshore technology and sea pipelines; use of computers in gas industry operations; gas use and distribution; gas heat pumps; efficiency of gas-using applicances and equipment; use of gas as a motor-vehicle fuel; gas statistics;

(c) <u>User-specific activities</u>

- <u>Industry</u>: increased use of coal in industry and district heating; electro-technology applied to industrial processes; use of electric heat pumps; use of electric energy in the chemical industry.

- <u>Households</u>: long-term forecasting of the utilization of electric power and heat in agricultural production and households; use of electric heat pumps, gas heat pumps; hot water production and space heating on the basis of combined energy sources (electricity, fossil fuels); impact of energy considerations on human settlements and land-use planning.

- Evaluated the impact of energy developments in the ECE region on the world, including developing countries, and vice-versa;

- Contributed to the optimization of energy, environmental and macro-economic policies.

7. On average, every year 20 meetings are held at the policy and expert level as well as 3-4 seminars, to which a wider interested public has access; 15-20 studies, surveys or seminar reports are concluded and released to the public domain. Annual Bulletins are published in the fields of coal, gas, electric power and general energy.

F. DOCUMENTATION AND INFORMATION

8. A list of available documents is given overleaf. Copies and further information can be obtained from the Energy Division, Economic Commission for Europe, 8-14, avenue de la Paix, 1211 Geneva 10.

ECE COMMITTEES SPECIALIZING IN THE FIELD OF ENERGY

COMMISSION

COAL COMMITTEE	COMMITTEE ON ELECTRIC POWER	COMMITTEE ON GAS	SENIOR ADVISERS TO ECE GOVERNMENTS ON ENERGY (ad hoc body)

COAL COMMITTEE

— Working Party on Coal Trade and Statistics

— Meeting of Experts on Opencast Mines

— Meeting of Experts on the Utilization and Preparation of Solid Fuels

— Meeting of Experts on Productivity and Management Problems in the Coal Industry

— Meeting of Directors of National Mining Research Institutes

COMMITTEE ON ELECTRIC POWER

— Meeting of Experts on Problems of Planning and Operating Large Power Systems

— Meeting of Experts on Electric Power Stations

— Meeting of Experts on the Relationship between Electricity and the Environment

COMMITTEE ON GAS

— Meeting of Experts on the Use and Distribution of Gas

— Meeting of Experts on the Transport and Storage of Gas

— Meeting of Experts on Natural Gas Resources

— Meeting of Experts on Gas Statistics (ad hoc)

AVAILABLE DOCUMENTS ON ENERGY

A.   PUBLICATIONS

Coal

International Codification System of Medium and High Rank Coals,
United Nations, New York 1988, ECE/COAL/115 (Sales No. E.88.II.E.15).

General Energy

Energy Efficiency in European Industry, ECE Energy Series
(Sales No. GV.E.88.0.8).

Environment

Air Pollution Studies, No. 9: Effects and control of Transboundary Air
Pollution (ECE/EB.AIR/13) (Sales No. 87.II.E.36).

B.   STUDIES

Coal

The coal situation in the ECE region in 1987 and prospects for
coal - Highlights (COAL/R.137).

Environmental problems resulting from coal mining and ancillary
activities - List of projects relating to the environment protection
(COAL/R.141).

World coal trade up to the year 2000 - Consumption of coke in the steel
industry (COAL/WP.1/R.101).

Statistics on the transformation of coal into secondary energy sources
(COAL/WP.1/R.102).

International standardization of methods of assessing shiploads and of
the relating documents and certificates (COAL/WP.1/R.104).

Capital investment and production costs in the ECE coal industry, 1983 up
to and including 1986 - Consolidated statistical tables (COAL/WP.1/R.105).

Consideration of factors influencing labour productivity by taking high
productivity mines and faces as examples, in conditions of thin and steep
seam inclinations, both in hard and brown coal undergound mining
(COAL/GE.1/R.71).

Research and development work in the field of brown coal/lignite
preparation and briquetting (COAL/GE.3/R.75).

Fluidized-bed combusion of low-calorific-value fuels (COAL/GE.3/R.78).

Exchange of experience on modern technologies relating to close water
circuits and reduction of moisture content during the preparation of
fines and slurries (COAL/GE.3/R.81).

Problems encountered in the establishment of plant growth on infertile overburden (COAL/GE.5/R.36).

Water management and protection in opencast mining regions and utilization of mine water (COAL/GE.5/R.37)

Overcoming the environmental problems in opencast mining (COAL/GE.5/R.46).

Electric Power

The electric power situation in 1987 and its prospects (EP/R.140 and Adds.1 and 2).

Applications and implications of high temperature superconductors (EP/R.145).

Heat pumps in agriculture and horticulture (EP/GE.1/R.51).

Problems arising from the design of high-capacity conventional and nuclear power stations in seismic areas (EP/GE.3/R.96).

Problems arising from the design, construction and operation of highly-manoeuvrable thermal power units that supply the half-peak of the load curve (EP/GE.3/R.97).

Gas

Information on institutes concerned with nitrogen oxide reduction and research (GAS/R.153).

Draft declaration on the contribution of gas to the reduction of environmental nuisances and on recommended measures (GAS/R.159).

The gas situation in the ECE region in 1987 and its prospects - Highlights (GAS/R.162).

Impact of new technology on the use of gas where it has a premium value (GAS/GE.2/R.96).

Use of natural gas and/or LPG to replace other fuels, mainly oil (GAS/GE.2/R.97).

Gasification of soild fuels and wastes (GAS/GE.2/R.100).

Use of gas heat pumps (GAS/GE.2/R.104).

Measures of gas conservation and study of results (GAS/GE.2/R.105).

Use of compressed natural gas as motor-vehicle fuel (GAS/GE.2/R.107).

Utilization of LPG as automotive fuel (GAS/GE.2/R.109).

Development of underground gas storage facilities (GAS/GE.3/R.77).

Gas pipelines transmitting gas from different owners and/or countries (GAS/GE.3/R.85).

Technical and economic aspects of major repairs to gas pipelines (GAS/GE.3/R.87).

General Energy

Energy issues and co-operation in the ECE region (ENERGY/R.43).

East-west energy trade in 1987 (COAL/R.135; EP/R.138; GAS/R.158; ENERGY/R.49).

Convention of Long-range Transboundary Air Pollution

Cost-impact and economic impact analyses of different SO_x and NO_x abatement strategies (EB.AIR/GE.2/R.18).

Annual review of strategies and policies for air pollution abatement (EB.AIR/R.32).

Agriculture

Situation and future use of new and renewable sources of energy in agriculture (FAO/ECE/AGRI/WP.2/R.120).

C.   SEMINARS

Housing, Building and Planning

Seminar on Policies for Energy Conservation in Buildings, Espoo (Finland), 6-10 June 1988 (HBP/SEM.40/2).

Chemical Industry

Seminar on the use of Electric Energy in the Chemical Industry, Lyon (France), 7-10 March 1988 (CHEM/SEM.17/2).

Electric Power

Seminar on the Impact on Atmospheric Protection Measures on Thermal Power Stations, Essen (Federal Republic of Germany), 19-21 September 1988.

D.   REPORTS OF MAJOR MEETINGS

Coal

Eighty-fourth session of the Coal Committee (26-29 September 1988 - ECE/COAL/117).

Electric Power

Forty-sixth session of the Committee on Electric Power (25-28 January 1988 - ECE/EP/76).

Gas

Thirty-fourth session of the Committee on Gas (18-21 January 1988 -
ECE/GAS/92).

General Energy

Sixth session of the Senior Advisers to ECE Governments on Energy
(24-27 May 1988 - ECE/ENERGY/13).

E.   STATISTICAL AND DATA BASE

Coal

Annual Bulletin of Coal Statistics for Europe 1980, 1984, 1985, 1986,
1987, Vol. XXII 1987, New York, 1988 (Sales No. E/F/R.88.II.E.6).

Electric Power

Annual Bulletin of Electric Energy Statistics for Europe 1980,
1984, 1985, 1986, 1987, vol. XXXIII 1987, New York, 1988
(Sales No. E/F/R.88.II.E.8).

Gas

Annual Bulletin of Gas Statistics for Europe 1980, 1986, 1987,
Vol. XXXIII 1987, New York, 1988 (Sales No. E/F/R.88.II.E.7).

General Energy

Annual Bulletin of General Energy Statistics for Europe 1985, 1986,
Vol. XIX 1986, New York, 1988 (Sales No. E/F/R.88.II.E.5).

Printed at United Nations, Geneva   06000P   United Nations publication
GE.91-30023   Sales No. E.91.II.E.2
March 1991—3,350—

ISBN 92-1-116499-0